Engineering G

GEOSCIENCE TEXTS

SERIES EDITOR

A. HALLAM
Lapworth Professor of Geology
University of Birmingham

1 Principles of Mineral Behaviour
A. PUTNIS and J. D. C. MCCONNELL

2 An Introduction to Ore Geology
A. M. EVANS

3 Sedimentary Petrology: an Introduction
MAURICE E. TUCKER

4 Geophysical Surveying
P. KEARY and M. BROOKS

5 Engineering Geology
F. C. BEAVIS

GEOSCIENCE TEXTS VOLUME 5

Engineering Geology

F. C. BEAVIS
MA, BSc, PhD, FGS
Professor of Engineering Geology
University of New South Wales

BLACKWELL SCIENTIFIC PUBLICATIONS
MELBOURNE OXFORD LONDON EDINBURGH BOSTON PALO ALTO

To Joan and To Evan

Editorial offices:
107 Barry Street, Carlton
Victoria 3053, Australia
Osney Mead, Oxford OX2 0EL, UK
8 John Street, London WC1N 2ES, UK
23 Ainslie Place, Edinburgh
EH3 6AJ, UK
52 Beacon Street, Boston
Mass. 02108, USA
744 Cowper Street, Palo Alto,
Ca. 94301, USA

First published 1985

Typset in Australia by Comset Pty Ltd.
Printed in Singapore by
Kyodo Shing Loong Printing Ind. Pte Ltd.

DISTRIBUTORS

USA and Canada
Blackwell Scientific
Publications Inc
PO Box 50009, Palo Alto
Ca. 94303

Australia
Blackwell Scientific Book
Distributors
31 Advantage Road, Highett
Victoria 3190

South-East Asia
P G Publishing Pte Ltd
Alexandra PO Box 318
Singapore 9115

Others
Blackwell Scientific
Publications
Osney Mead, Oxford
OX2 0EL, UK

Cataloguing in publication data
Beavis, F.C. (Francis Clifford), 1924–.
Engineering Geology.

1. Engineering geology. I. Title. (Series: Geoscience texts; v.5).

Bibliography.
Includes index.
ISBN 0 86793 200 7.
ISBN 0 86793 128 0 (pbk.).

624.1'51

Contents

Preface

Introductory textbooks on geology for civil engineering students quite rightly place the emphasis on geology. The applications of geology to engineering, however, are often divorced from the basic geology, and the student, in consequence, sometimes fails to comprehend as well as he should the close relationship between the two disciplines. In this book, the aim has been as close an integration as possible. Discussion of geological principles has been linked with the application of these principles to engineering practice. Illustrative examples, drawn from a wide geographic range, are used freely. The case histories were chosen from both failures and successes, for it is from both, not the failures alone, that we learn. So far as has been possible, the case histories selected are projects with which the writer has had some direct personal contact. The two final chapters collect, in a systematic form, the main geological factors involved in the safe and economic construction of engineering works, and some applications of geology to environmental studies.

An emphasis is placed, on the geological side, on those aspects of geology of importance to the engineer: mineralogy, petrology, structural geology, geomorphology and rock weathering. No attempt has been made to include more than the basic principles of stratigraphy. Detailed discussion of local stratigraphic sequences is of little general interest, or value, to the engineering student, although, of course, on an actual project knowledge of the local stratigraphy is always essential for the geologist, and at least helpful to the engineer.

In preparing the book, the author has kept in mind the needs of middle-level civil engineering students, and a knowledge of secondary school geology has been assumed. Reference lists have been largely restricted to suggestions for further reading from more advanced standard texts. However, some references to original literature are given to encourage the student to obtain more detail, and to gain experience in critical evaluation of original work.

The author wishes to acknowledge his gratitude to copyright holders for permission to use published material. Specific acknowledgement is made in the text. Many people have contributed to the preparation of the book, and to all of them, I am sincerely grateful. My wife, Joan, has always given her support and encouragement. Mrs Diane Blank carefully and diligently prepared the typescript, Mrs Marianna Horvath the line drawings, and Mr G. Small the half tone plates. Mrs Blank and Mrs Larisa Smith prepared the indexes. I am particularly grateful to my colleague, Dr M.J. Knight, who provided much of the information and many of the illustrations contained in Chapter 10. His knowledge of environmental geology, and his willingness to discuss this aspect with me, have been invaluable. Finally, I wish to acknowledge the encouragement I received many years ago from Mr E.L. Richard, to whom this book is jointly dedicated.

Frank Beavis
Sydney

1

Introduction

Engineering geology, as a science, is concerned with the applications of the principles of geology to civil (and to some extent, mining) engineering, so that the engineer can take into account those geological aspects which control the economy, and safety, of the structure which he is required to design and construct. All civil engineering works are constructed on, or in, rock and soil masses. It is essential, therefore, that the civil engineer be aware of the history, nature, and properties of the rocks and soils. Equally, it is essential for the geologist, who is to work with the engineer, to have some knowledge of engineering requirements.

Although engineering geology has been practised for centuries, systematic study began only in the 19th century. The present century saw the development of soil mechanics and rock mechanics; this has tended towards a so-called 'quantification' of engineering geology. Such a quantitative approach is desirable, but it must not be overlooked that much of the geology which is so significant for the engineer is qualitative and descriptive. Rock mechanics and soil mechanics supplement, or complement, the descriptive geology, they do not replace it.

At the most fundamental level, geological principles must be used to explain the variation in the mechanical properties of rocks and soils, and the behaviour of rocks and soils under stress. The quantitative description of soil and rock behaviour has to assume an idealized material; the geologist is aware that actual soil and rock materials, in their texture, structure, and composition vary to a considerable degree from the ideal.

The science of geology comprises a number of subdisciplines, all of which, to a greater or lesser extent, find some application in engineering. An understanding of geology itself requires a sound basis in chemistry, physics and mathematics, since geology is concerned with processes which are chemical and physical, and with the products of the various processes.

Mineralogy is that branch of geology which is concerned with the origin, occurrence, and the properties of minerals. Any mineral species is a unique combination of a definite chemical composition and crystalline structure. Several thousands of mineral species are known, but the civil engineer, for the most part, is concerned with only a few: those which are known as rock-forming minerals. Some minerals are of particular interest and importance to, especially, concrete engineers, since they react with cement under certain conditions; the reaction can result in the disruption of concrete.

The study of the occurrence, distribution, origin, composition and texture of rocks is known as petrology. All rocks are made up of one or more minerals. The minerals which are present, their arrangement (fabric) and the texture, are determined by the origin of the rock and the processes it has undergone, and is undergoing, since its formation. Rocks can be classified genetically into igneous, sedimentary and metamorphic, depending on their origin. Rocks in each of these classes have certain, well defined characteristics, which are reflected in their behaviour as engineering materials.

As a result of the processes to which rocks are subjected during and after their formation, structural features such as bedding planes, joints, faults, folds and foliations may be present. The origin, geometric patterns, and the forces which produced these structures constitute the subdiscipline of structural geology. Most structures represent discontinuities in the rock mass, while some represent discontinuities in the rock material. As a result they have a major influence on the engineering properties, and behaviour, of both mass and material. Structural geology is one of the most important basic studies for engineering geology.

Although of less interest to the engineer, those branches of geology known as palaeontology and stratigraphy are of vital importance to the geologist, and are used, quite commonly, to solve problems which confront the engineer. Palaeontology is the science of fossils: the remains of past life on the earth. Fossils are used particularly to determine the age of the rocks in which they are found. Other methods of dating are available which give absolute ages, as distinct from the relative ages determined by means of fossils. By studying the age, the structure, and the field relationships of rock masses, geologists are able to determine their history, and, in the final analysis, the history of the whole earth. The basis of any geological map is the stratigraphy of the region involved. The history of the region can have consequences of vital importance to engineering.

The final stage in the geological evolution of any region is the development of the present land surfaces. The processes responsible for this development are largely climatically controlled, although the landforms themselves are functions both of climate (present and past), and of underlying geology. Geomorphology is the study of the form and origin of the present land surface. Engineering geomorphology is emerging as a specialist discipline, particularly in the fields of slope engineering and enviromental engineering. Nonetheless, a careful geomorphological appraisal of an engineering site has always been an essential facet of an investigation programme.

Apart from the somewhat purely geological subdisciplines mentioned, engineering geological investigations require, in addition, studies which overlap with engineering: soil mechanics, rock mechanics, and groundwater hydrology. Hydrogeology is concerned with the origin, quality, quantity, distribution and movement of water in, and through, a rock mass. Although hydrogeologists are interested primarily in groundwater as a useable resource, engineering geologists are concerned with the hydrology of groundwater at most engineering sites, but especially at dam sites and in tunnels and other underground excavations.

It has been noted that the basic data relating to the geology of a region are recorded on a geological map. While such maps are indispensable to the engineer, not infrequently they fail to show many of the factors of importance for design and construction. In recent years, the basic geological map has come to be supplemented by specialized maps recording specifically engineering geological features. These are referred to as engineering geological maps, engineering geomorphic maps, and geotechnical maps. These are all prepared after field, laboratory and office investigation and research.

In the following pages, the basic principles of geology are outlined. The applications of geology to civil engineering are presented, and examples of these applications are discussed. The reader who is concerned that the treatment is for the most part qualitative should note the words of Sir Harold Harding, a former President of the Institution of Engineers, London: "Some engineers despise geology because it is not quantitative, there are no formulae attached to it. But to me, geology is the basis of civil engineering . . ."

2

Rock — The Material

2. 1 General statement

Most rocks are aggregates of mineral particles which are crystals, more or less perfectly formed, or fragments of crystals; some rocks may contain, or be entirely composed of, non-crystalline materials such as natural volcanic glass. The size of the particles, their arrangement, and the proportion of each mineral present, serve as the basis for the geological classification; the origin and history of the rock are of major importance also since, to a large extent, these aspects will have determined the size, arrangement and composition of the mineral particles. Rocks are natural materials commonly lacking the uniformity of man-made materials such as steel. While it is possible for man to control the properties of manufactured engineering materials, he can exert only a very limited control, if any, over the properties of rocks and soils.

In engineering geology, it is essential to distinguish between the rock or soil *material* and the rock or soil *mass*. Because of textural and compositional variations, any sample of rock or soil is almost certainly not the same, in all respects, as the mass from which it was taken. Apart from textural and compositional variations, the mass may be broken by discontinuities such as joints or bedding planes which influence the engineering properties, and the behaviour under stress, of the rock mass. For these, and for other reasons which will become apparent, it cannot be emphasized too much that the value of properties determined on a small sample of rock or soil may bear little relationship to the values for the mass. In engineering geology it is essential that a clear distinction be made between material and mass characteristics.

2. 2 Rock-forming minerals

A mineral is a naturally occurring crystalline material with a definite chemical composition and a definite crystal structure. Many thousands of mineral species have been recognized, but only a few of these commonly occur as rock-forming minerals. Most of the rock-forming minerals are silicates (Table 2.1). Each mineral in a rock has physical, mechanical and chemical properties which differ from those of the other minerals present, and which will have some effect on the properties of the rock as a whole. However, it is generally true that, since the individual mineral particles in a rock are small, usually each particle can have, of itself, little direct influence on the mechanical properties of the rock as a whole. Properties such as hardness, or a well-developed mineral cleavage, in a mineral which forms a significant part of the rock can, under some conditions, determine the reaction of the rock to external forces. For example, the preferred orientation of mica crystals, each with a highly developed platy cleavage, can impart a strong mechanical anisotropy to a rock such as slate, phyllite or schist.

Table 2.1 Common rock forming minerals.

Family	Species	Structural class	Composition
	Quartz	Tektosilicate	SiO_2
Feldspar	Orthoclase Plagioclase	Tektosilicate	$KAlSi_3O_3$ $NaAlSi_3O_8$–$CaAlSi_2O_8$
Mica	Muscovite Biotite	Phyllosilicate	$KAl_2(AlSi_3O_{10})-(OH)_2$ $K_2(MgFe)_6(SiAl)_8O_{20}(OH)_4$
Amphibole	Hornblende	Inosilicate	$(NaCa)_2(MgFeAl)_5(SiAl)_8O_{22}(OH)_2$
Pyroxene	Augite	Inosilicate	$Ca(MgFeAl)(AlSi)_2O_6$
Olivine	Olivine	Nesosilicate	$(MgFe)_2SiO_4$
Clays	Kaolinite Illite Montmorillonite	Phyllosilicate	$Al_4Si_4O_{10}(OH)_8$ $KAl_2(AlSi_3)O_{10}(OH)_2$ $Al_2Si_4O_{10}(OH)_2nH_2O$
Carbonates	Calcite Dolomite		$CaCO_3$ $CaMg(CO_3)_2$

2.2.1 SILICATE MINERALS

The silicate minerals consist essentially of silicon–oxygen tetrahedra, linked together in a variety of ways; silicate mineral classification is based on the type of linkage between the tetrahedra. Six structural classes are recognized, as shown in Table 2.2.

Table 2.2 Silicate mineral structural classes.

Structural class	Structure	Example
Nesosilicate	Independent tetrahedra	Olivines
Sorosilicate	Double tetrahedra	Epidote
Cyclosilicate	Ring structure	Cordierite
Inosilicate	Chain structure	
	single chain	Pyroxenes
	double chain	Amphiboles
Phyllosilicate	Sheet structure	Micas, clays
Tektosilicate	Three-dimensional network	Quartz, feldspars

The other components of the silicate structures such as cations, additional oxygen, hydroxyl groups, and water, are arranged within the structure in such a way as to produce physical stability and electrical neutrality. The *cyclosilicates* and *sorosilicates* do not include any really important rock-forming minerals although epidote and cordierite may be essential constituents of some metamorphic rocks. On the other hand, the *tektosilicates* include some of the most important rock-forming minerals: quartz, feldspars, feldspathoids and zeolites. Of all minerals, quartz and feldspars are the most abundant in rocks. All of the tektosilicate minerals are colourless, white, or grey, provided they are free from inclusions or impurities. As a result of the relatively open framework, their density tends to be low, and their hardness is uniform between 5 and 7. On the whole, the tektosilicates form an extremely homogeneous group. *Phyllosilicates* are noted for the highly developed platy basal cleavage. The main minerals are the micas, the clays, pyrophyllite, talc, serpentine and chlorite. Although some phyllosilicates are stable to quite high temperatures, many of them are formed at low temperatures during weathering and sedimentary processes. The chain structures of the *inosilicates* are responsible for the well-defined prismatic cleavages characteristic of the amphiboles and pyroxenes. All of the minerals in this class have strikingly similar properties. The *nesosilicates,* which include garnets and olivine, have a very dense packing, which is reflected in the relatively greater hardness and higher density than is found in corresponding compounds of other structural classes, while the absence of chains and sheets is reflected in the generally equidimensional nature of the crystals.

2.2.2 CLAY MINERALS

The use of the term *clay* to describe a material implies that it has a very fine texture, with individual particles less than 0.002 mm in diameter, and that it develops a plasticity when mixed with water. Clay, in this sense, is not composed exclusively of clay minerals, but these do constitute, as a rule, a significant proportion of the material, and it is certainly the clay minerals which are responsible for the plasticity of the clay. Clay minerals are of major importance to the engineer because of the influence that they exert on the mechanical properties of the soils and rocks in which they occur. All clay minerals are phyllosilicates (Figs 2.1, 2.2).

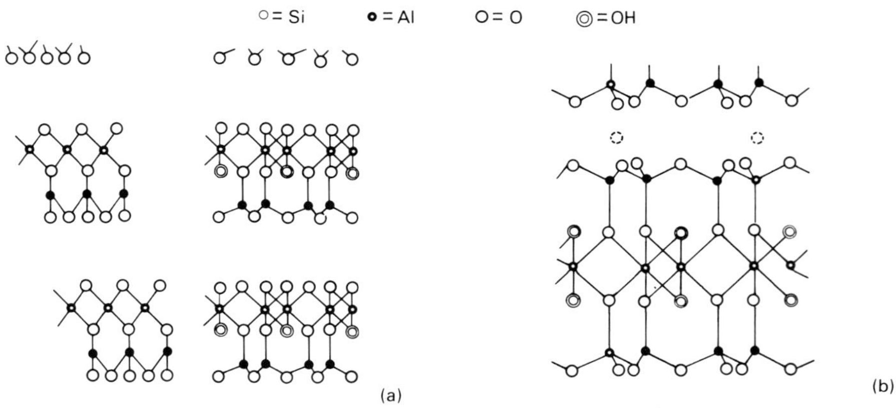

Fig. 2.1 The structure of clay minerals: (a) Kaolinite. (b) Illite.

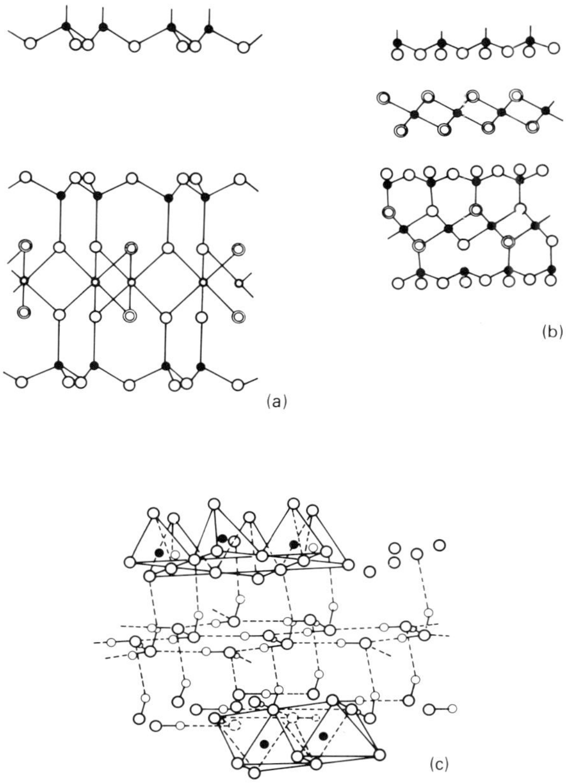

Fig. 2.2 The structure of clay minerals: (a) Montmorillonite. (b) Chlorite. (c) Hydrated halloysite.

Because of their economic importance, clay minerals have been studied in great detail; much is known of their structure and physico-chemical properties. While the clay minerals can be classified in a number of ways, a simple classification based on structure and physico-chemical properties, which is of value to the engineer, is presented in Table 2.3. In this classification, five main classes are recognized.

The 1:1 type clay mineral crystal units are made up of one tetrahedral silica layer alternating with one octahedral alumina layer. The two sheets of each unit are held together by oxygen atoms which are shared by the silicon and aluminium atoms in their respective layers. The units themselves are held together by relatively rigid hydrogen bonding. As a consequence, the lattice is fixed, and no expansion will normally occur when water is added. Neither water nor cations can enter between the structural units, so that kaolin, and other minerals in this type, have a low cation adsorption capacity. The plasticity, cohesion, shrinkage and swelling characteristics are low.

The 2:1 type expanding clays have crystal units characterized by an octahedral alumina layer confined between two tetrahedral silica layers. The crystal units are loosely bonded by very weak oxygen to oxygen links. Water molecules and cations are attracted into the inter-unit space, causing expansion of the crystal lattice. The movement of water and cations in between units exposes a large internal surface area in contrast to the 1:1 type clays. Isomorphous substitution of Mg^{2+} for some of the Al^{3+} in the octahedral layer and, to a lesser degree of Al^{3+} for Si^{4+} in the tetrahedral layer leaves a mineral such as montmorillonite with a high net negative charge, so that montmorillonite has a high cation absorption capacity: up to 15 times that of kaolinite. The montmorillonite minerals have a high plasticity and cohesion, and a high shrinkage on drying.

Table 2.3 Classification of clay minerals.

A. 1 : 1 Type		One silica tetrahedral layer alternating with one alumina octahedral layer
		Kaolinite
		Halloysite
		Anauxite
		Dickite
B. 2 : 1 Type Expanding		An alumina octahedral layer contained between two silica tetrahedral layers
	B(1)	Montmorillonite Group
		Montmorillonite
		Beidellite
		Nontronite
		Saponite
	B(2)	Vermiculite
C. 2 : 1 Type non-expanding		
		Illite
D. 2 : 2 Type		Two silica tetrahedral sheets and two magnesium octahedral sheets
Chlorites		
E. Mixed layer clays		
F. Fibrous clays		
		Palygorskite
		Sepiolite

The vermiculites have stuctural characteristics similar to those of the montmorillonites, but in some the octahedral layer is dominated by magnesium rather than aluminium. Water molecules and magnesium ions are strongly adsorbed between the crystal units. However, these tend to hold the units together, so that the degree of swelling is less than in the montmorillonites. The cation adsorption capacity of vermiculite exceeds that of all other clay minerals, due to the high negative charge in the tetrahedral layer.

The 2:1 non-expanding clay minerals are sometimes referred to as hydrous micas. Illite is the most important member of the type. In this type, some 15% of the tetrahedral silicon sites are occupied by aluminium atoms, which results in a high negative charge in the tetrahedral layer. To satisfy this charge, potassium ions are strongly attracted between the units. The ionic radius of the potassium is the same as spaces in the adjacent tetrahedral layers, so that the potassium acts as a bonding agent, preventing expansion of the crystal lattice. As a result, properties such as hydration, cation adsorption, swelling, shrinkage, and plasticity are less in illite than in montmorillonite. Nonetheless, illite has these properties in excess of the 1:1 clay minerals.

The 2:2 type minerals are represented by chlorite which is composed of alternate talc and brucite [$Mg(OH)_4$] layers. Thus the crystal unit consists of two silica tetrahedral sheets and two magnesium octahedral sheets. The cation exchange capacity is about the same as illite. Relatively little water absorption occurs between the layers, so that this group of minerals is relatively non-expansive.

A very important group of clay minerals is the co-called 'mixed layer' or 'interstratified' clays. Within a given crystal, the crystal units may be of more than one type, e.g. illite–montmorillonite. In some soils and weathered rocks, mixed layer clays may be more common than single structure types.

The type and amount of clay mineral present in a rock, but particularly in a soil, can have a profound effect on engineering properties. The nature of the exchangeable ion, and the ion-exchange capacity of the clay mineral are also important to the engineer, especially in soil stabilization processes.

2.2.3 CARBONATE MINERALS

Many carbonate minerals exist, some of which are important as metallic ores. Only two, however, are of real importance as rock-forming minerals; calcite and dolomite, although others, such as siderite, $FeCO_3$, magnesite, $MgCO_3$, and aragonite, $CaCO_3$, may be important as rock constituents. The fundamental anionic unit of the carbonates, CO_3^{2-}, is a planar structure, having a carbon atom at the centre of an equilateral triangle, with an oxygen atom at each of the apices. The structure is such that solution can occur readily, although $CaCO_3$ is insoluble under alkaline conditions, and in sea water. Many of the carbonates, including calcite and dolomite, are relatively soluble in slightly acidic water. Dolomite is markedly less readily soluble than calcite.

Calcite is the principal, sometimes the sole, constituent of the rocks limestone and marble, while dolomite is an important constituent of the dolomitic marbles and dolomitic limestone, and is the principal mineral in the rock dolomite. Both calcite and dolomite have a strong rhombohedral cleavage and both develop twin lamellae as strain phenomena when subjected to stress. Both of these characteristics are important in determining the mechanical properties of the minerals.

2.3 Geological classification of rocks

Rocks may be classified in a number of ways, depending on the purpose for which the classification is required. In a geological classification, the main considerations are the mineral composition of the rock, the physical and chemical interaction between the mineral grains, and the processes which have affected the rock during and after its formation. In an engineering classification of rock material, emphasis is placed on those aspects which influence engineering behaviour. Geological classifications have only a very limited application in engineering geology; they cannot be ignored since often an engineering meaning is attached to a rock name, and since they do highlight some features of a rock which can be important to the engineer. These include the processes involved in rock formation, the mineral composition, texture and fabric. Fundamentally, the main geological classes are genetic, so that rocks are classified according to origin as igneous, sedimentary or metamorphic. Finer subdivision is based on mineral composition; texture, which refers to the size and shape of grains; and fabric, which refers to the mutual arrangement of grains. Geologists recognize thousands of different rock types based on differences, often slight, in composition, texture and fabric. In engineering geology many of these rocks can be grouped together without any serious consequences.

2.3.1 IGNEOUS ROCKS

Rocks included in this genetic class are those which have cooled and crystallized from a molten material called *magma.* The type of rock produced from the magma depends on the cooling history, during which a number of physical and chemical processes, known as differentiation, occur; and on the composition and environment of the magma. Sometimes the magma cools and crystallizes before

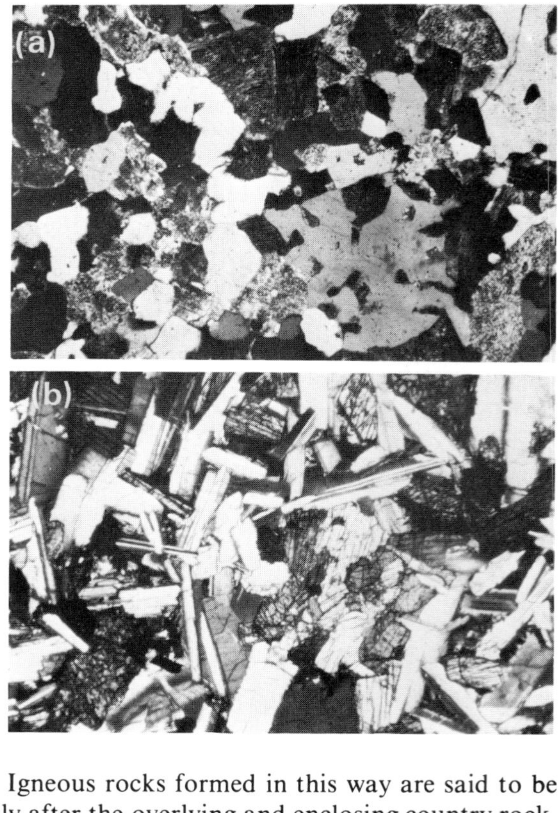

Fig. 2.3 Igneous rocks: (1) Granite, Yalwal Creek, Australia (×36). (2) Basalt, Mt Bogong, Australia (×36).

it reaches the earth's surface. Igneous rocks formed in this way are said to be *plutonic;* they are exposed only after the overlying and enclosing country rock, into which they have been intruded, has been eroded away. Because of the insulating effect of the enveloping country rock, often many kilometres thick, plutonic rocks cool and crystallize over a long period of time, perhaps millions of years. As a result, these rocks have a coarse texture, with individual grains reaching relatively large sizes. If the magma breaks through to the surface, a volcano is developed, and the rocks are said to be *volcanic*. Cooling and solidification occur very rapidly: often over a period to be measured in hours, days or weeks. Because of the rapidity of cooling, volcanic rocks have a very fine texture and are sometimes even glassy, the cooling having been too rapid for crystallization to take place. Under some circumstances, such as explosive eruptions, the volcanic rocks may be fragmental tuffs and agglomerates. In any case, lava flows possess a broken base and a broken upper surface due to the rapid formation of a crust on the flow. Since igneous rocks, both plutonic and volcanic, have been in a liquid state and have flowed, a flow fabric may be present, expressed as a preferred dimensional orientation of platy and elongated mineral grains, and in volcanic rocks, of gas vesicles. This flow fabric has the effect of producing a planar or linear mechanical anisotropy in the rock.

Finer subdivision of igneous rocks is based on mineral composition which is a reflection of rock chemistry. Four groups are generally recognized: acid, intermediate, basic and ultrabasic. Acid rocks, with a silica content of 66% or more, contain free quartz in relative abundance; in the intermediate rocks, with SiO_2 between 54% and 66%, quartz is a minor constituent, with the feldspars dominant. Basic rocks have SiO_2 contents between 45% and 54%. Quartz is absent, and, while feldspars are present, they may be dominated by dark minerals, but over 50% feldspar is common. Ultrabasic rocks, with SiO_2 contents less than 45%, are composed almost exclusively of the dark minerals. Many geological classifications of igneous rocks have been developed. In engineering geology, a simple classification such as that given in Table 2.4 is quite adequate. This classification is due to the Geological Society Engineering Group Working Party on the Description of Rock Masses for Engineering Purposes, and should be adopted as a standard. The classification is based on simple criteria which can be determined readily in the field: an essential requirement.

2.3.2 SEDIMENTARY ROCKS

All sedimentary rocks owe their origin to a long series of processes, the last of which are deposition and the conversion, by pressure and chemical reactions, of the loosely deposited sediments into a compact rock. Deposition of the materials from which sedimentary rocks are formed usually occurs in layers or lenses which are known as beds, or strata. For this reason, all sedimentary rocks possess, to a greater or lesser degree, a planar mechanical anistropy. Bedding is one of the most characteristic features of this group of rocks. Many kinds of sedimentary rocks exist, but they can be grouped into three main classes: detrital or clastic; volcanic-clastic and pyroclastic; and chemical-organic.

Clastic sedimentary rocks have been derived from other, pre-existing rocks by weathering; erosion; and transportation of the products of erosion, by wind, water, or ice; and finally diagenesis, the physical and chemical processes which

Table 2.4 Geological classification of igneous rocks.

Structure	Texture	Group	Acid Quartz, feldspar and dark minerals	Intermediate	Basic Feldspar and dark minerals	Ultrabasic Dark minerals
Usually massive	Very coarse	Plutonic	Pegmatite			Pyroxenite
	Coarse		Granite	Diorite Syenite	Gabbro	Peridotite
	Medium		Microgranite	Microdiorite Microsyenite	Dolerite	
Usually as flows	Fine to very fine	Volcanic	Rhyolite	Trachyte Andesite	Basalt	Picrite Limburgite
	Glassy		Obsidian Pitchstone Pumice		Tachylyte	

convert a loose unconsolidated sediment into a consolidated rock. Deposition may be subaerial or subaqueous, and this factor, together with the mode of transportation, will influence the fabric, and to some extent the composition, of the rock, and hence its mechanical properties. Diagenetic processes may be physical or chemical, or both. The main physical process is compaction which produces a reduction in porosity and moisture content. The chemical processes are complex, but the result is the cementing, or welding, together of grains, and chemical reactions within and between grains which produce new minerals. The nature of diagenesis is of vital importance in engineering geology. If diagenesis was purely physical compaction, the rocks are much less stable physically and chemically than those in which chemical cementation and reaction have taken place.

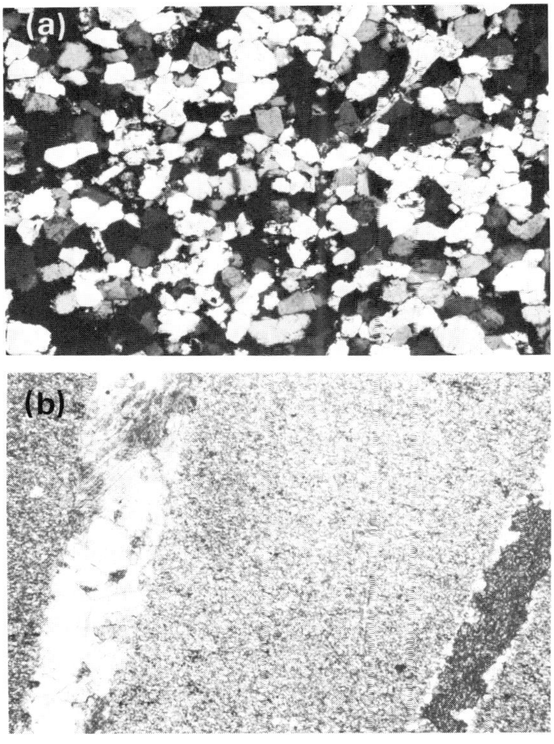

Fig. 2.4 Sedimentary rocks: (1) Sandstone, Fowlers Gap, Australia (×36). (2) Siltstone, Fowlers Gap, Australia (×36). Note the inhomogeneity and discontinuities.

Compaction sediments, and especially the finer textured types, tend to disintegrate when subjected to wetting, or to alternate wetting and drying; they may also show excessive swelling and shrinking, and abnormally high creep when the rock is loaded or unloaded. Unless the cement is highly soluble, cementation sediments do not show the extreme reactions to wetting and drying, nor do they exhibit the same degree of time-dependent strain when loaded or unloaded. Detrital sedimentary rocks for which diagenesis was essentially chemical are usually relatively stable both physically and chemically.

While detrital sedimentary rocks may be classified in purely physical terms such as grain size, the geological classification may take into account grain size, grain shape, fabric and mineral composition. As studies in sedimentary petrography advance, the classification of detrital sedimentary rocks is becoming increasingly intricate. The classification of these rocks, included in Table 2.5, uses those factors which can be determined simply, and readily, in the field, and which have some effect on the mechanical properties of the rock. In the arenaceous rocks, the rock name may be qualified by prefixing a term descriptive of the nature of the cement, or of the dominant mineral present. Unfortunately, this classification, due to the Geological Society Working Party, takes no account of the diagenetic processes.

Pyroclastic sediments, of which the most common examples are agglomerate and tuff, are the result of explosive volcanic activity and the associated lahars and mudflows. Mixtures of country rock, volcanic rock, and lava are thrown into the air and are then deposited, usually with a high moisture content due to rain and condensing steam, in layers. Each layer tends to represent a single explosive volcanic event.

Within the class of chemical-organic sediments is to be found a wide variety of rock types. The organic sediments have their origin in living organisms, both animal and plant. Plant remains ultimately become coal, while the calcareous skeletons of marine organisms accumulate to form limestones. Here, the diagenetic processes progressively cement the loose accumulation of skeletons and skeletal fragments to a compact crystalline rock, so that the term limestone covers a wide range of rock conditions, all exhibiting different mechanical properties. Since the original skeletons are calcareous, limestone is composed dominantly of calcite or aragonite. Because many limestones have a high proportion of detrital impurities, some workers place these rocks in a special class of detrital calcareous rocks. Although coal is normally regarded as a fuel, there are parts of the world where it has formed foundations for engineering structures. Little is known of its foundation characteristics, but it is a material being more closely studied in geomechanics as the stability of open-cut coal mines becomes a major problem requiring solution.

Chemical sediments are not very common, but large masses of these rocks do occur. They may result from the precipitation of salts from solution, by chemical reaction, or as a result of evaporation, or from biochemical activity. The most common chemical sediments are travertine, chert and flint, and cap rocks such as calcrete, silcrete and ferricrete.

2.3.3 METAMORPHIC ROCKS

Both the mineral assemblage which constitutes any rock, and the fabric of that rock, are stable within very narrow ranges of pressure and temperature. In the event of either, or both, of the pressure and temperature stability ranges being exceeded, changes occur in both rock composition and fabric in response to the new conditions; i.e. a new mineral assemblage, and fabric, are established which are stable under the newly imposed pressures and temperatures. The changes which occur constitute metamorphism; the rocks produced as a result of these

Table 2.5 Classification of sedimentary rocks.

Genetic group	Detrital	Detrital Calcareous	Pyroclastic	Chemical - non-calcareous Organic
Composition	Rock fragments, quartz, feldspar, other minerals	At least 50% of grains are calcite	At least 50% of grains are fine grained volcanic rock	
Very coarse grained to coarse grained >2 mm	Grains are rock fragments Rounded: Conglomerate Angular: Breccia	Calcirudite	Rounded: Agglomerate Angular: Volcanic breccia	
Medium grained 0.20–0.06 mm	Sandstone: grains mainly mineral fragments Quartzite: 95% quartz Arkose: 75% quartz 25% feldspar Greywacke: quartz, feldspar and rock fragments All angular	Calcarenite	Tuff	Halite Gypsum Dolomite Chert Flint Coal
Fine grained to very fine grained <0.06 mm	Mudstone Shale: fissile Siltstone: 50% fine grained particles Claystone: 50% very fine grained particles	Calcisiltite Calcilutite	Tuff	

changes are called metamorphic rocks. Rocks resulting primarily from elevated temperatures, as for example, around a plutonic intrusion, are known as thermal metamorphic rocks. When pressure is the main agent of metamorphism, dynamic metamorphic rocks are formed. This latter type of metamorphism is usually effective over a large volume of the crust, in major fold belts, while thermal metamorphism tends to be very localized. Usually, the effects of thermal metamorphism are compositional, with less obvious, but important changes in texture and fabric. On the other hand, dynamic metamorphism results not only in major changes in mineral assemblage, but also in quite dramatic changes in texture and fabric. It is the metamorphic fabric which is so important in determining the engineering properties of these rocks. Dynamic metamorphism results almost always in the mineral constituents developing a preferred dimensional orientation, and this imparts a planar anisotropy to the rocks, known as foliation, cleavage or schistosity. In mechanical terms, foliation is one of the most important characteristics of metamorphic rocks.

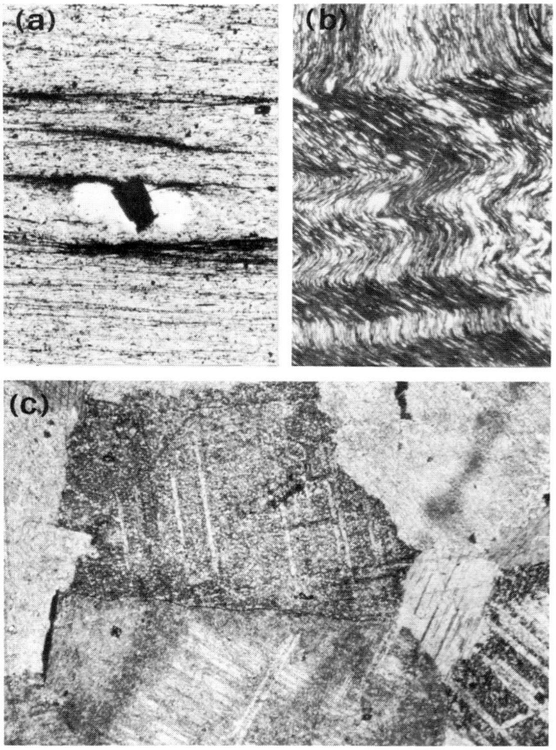

Fig. 2.5 Metamorphic rocks: (1) Slate, Bendigo, Australia (×36). Note the single planar anisotropic-slaty cleavage. (2) Chlorite schist, Tawonga, Australia (×36). Note two planar anisotropies, schisosity and strain slip cleavage. (3) Marble, Wombeyan, Australia (×36).

In those rocks which have undergone only a low grade of dynamic metamorphism, foliation is expressed as a fine slaty cleavage such as is seen in slate and phyllite. When the metamorphism has been more intense, the foliation is expressed as a coarse schistosity, often combined with a lithological layering resulting from the processes of metamorphic differentiation. In some areas, metamorphism and deformation may have occurred more than once, so that more than one foliation may have been imposed. Even in the simplest case of a single metamorphic event, two physical anisotropies may be present: bedding of the original sedimentary rock, and the imposed metamorphic foliation. For two metamorphic events, three planar anisotropies may be present, although with increasing complexity of metamorphism and deformation, some earlier foliations may be destroyed.

Table 2.6 is a classification of metamorphic rocks which is essentially geological, but which also includes criteria of importance to the engineer. The composition of the parent rock, the nature of the metamorphic processes and the intensity of metamorphism will determine the composition and fabric of the metamorphic rock. Metamorphic rock names are commonly prefixed by the names of the more prominent metamorphic minerals present, e.g. cordierite hornfels; almandine sillimanite schist.

Table 2.6 A classification of metamorphic rocks.

Fabric	Rock	Grain size	Composition
	Hornfels	Fine	
Granular	Metaquartzite	Fine to coarse	Quartz >70%
	Marble	Medium to coarse	Calcite >70%
	Granulite	Coarse	
	Slate	Very fine	Fine micas and clay minerals
Foliated	Phyllite	Very fine	Fine micas
	Schist	Medium	Quartz feldspar, micas
Foliated and	Layered schist	Medium	Quartz, feldspar, micas
layered	Gneiss	Coarse	Quartz, feldspar, micas

2.4 Rock fabric

As it is applied to rocks and soils, the term *fabric* is defined as covering the complete spatial and geometrical configuration of *all* of those components which make up the rock and soil. Within this definition is included texture and structure; i.e. the size and the preferred orientation of components, such as grains, crystal elements and structures, pores and voids, and the interrelationship between mineral grains. Put more simply, the fabric describes the shapes, sizes, orientations and any other relevant characteristics of the individual components of the rock or soil material (and mass). For purposes of statistical analysis of fabric, it is usual to regard as fabric elements only those features which are penetrative, i.e. which occur everywhere within the rock. Here, however, the scale of observation must be taken into account since an element which may be penetrative in terms of the rock mass, say joints and bedding planes, may not be penetrative in terms of the rock material. Some elements are penetrative irrespective of scale, for example, foliation in a metamorphic rock. It is necessary when discussing fabric always to define both the fabric element under discussion, and to specify the scale of observation.

The statistical analysis of fabric at all scales is important in the study of the response of rock to stress. Elements usually considered at a small scale, in terms of the rock material, are the preferred orientation of inequant crystals or grains, and the preferred orientation of microdefects such as microfractures, mineral cleavages, and strain twin lamellae. Under some special circumstances, crystallographic elements such as optic axes of minerals (e.g. quartz and calcite) may be analysed to determine the existence of a preferred orientation which could influence the mechanical behaviour of a rock, and which may otherwise not be observed. On a somewhat larger scale of a handspecimen or single outcrop, such elements as foliation and bedding may be examined, while on the largest scale, that of the rock mass, foliation, bedding, fold axes and joints will be examined. Fabric analysis at this scale and of these elements is essential in a geological engineering study of a rock mass.

The purpose of fabric analysis is to determine, in a quantitative manner, the existence and pattern of elements, often mechanical defects, which influence the isotropy and continuity of both the material and the mass. It will be shown later that anisotropy, and discontinuity, even at the smallest scale, can influence such important properties as strength and elasticity of a rock. Fabric analyses are useful, too, in the initial stages of the stress analysis of a rock mass. Some analyses can be performed only by specialists, e.g. clay mineral fabrics. Others, such as the analysis of foliations, bedding and joints is a simple task for any geologist or geotechnical engineer, especially if use is made of modern computer techniques.

Fabric data are usually analysed, and presented, in the form of contoured equal area diagrams (Fig. 2.6). The two most important characteristics of the fabric, the geometry of the elements and the symmetry, can be determined in this way. Figure 2.6 is actually an analysis of a rock mass fabric, but it serves well to illustrate geometry and symmetry. Poles to 1513 joints in a granitic gneiss (from north-eastern Victoria, Australia) were plotted, and concentrations contoured.

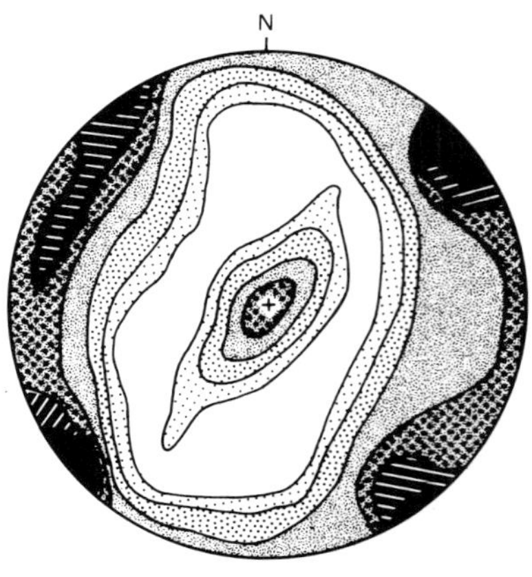

Fig. 2.6 Equal area projection of 1513 joints, Kiewa, Australia. Contours 6–5–4–3–2–1%.

It can be seen that there is a wide range in the orientation of the joints, but three concentrations exist corresponding to strike north-east and vertical dip; strike north-west and vertical dip; and horizontal. A definite pattern and an orthorhombic symmetry for this fabric are clearly established.

Fabric symmetry is important for both theoretical and practical studies in geomechanics. Figure 2.7 shows a series of fabrics of rock materials (orientations

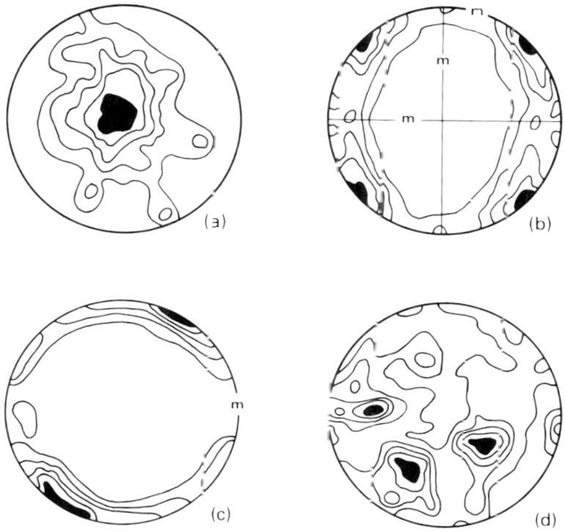

Fig. 2.7 Fabric symmetry: (a) Axial symmetry. (b) Orthorhombic symmetry. (c) Monoclinic symmetry. (d) Triclinic symmetry. [After Turner and Weiss (1963)]

of cleavages in mica) each with a different symmetry, the symmetry being determined by the existence of a centre, axes, and planes of symmetry. The basic principle, and the importance, of fabric symmetry lies in the fact that it is a reflection of the symmetry of both the deforming stress, or stresses, and of the movements which produced the fabric. It is possible, then, by determining the symmetry of the fabric, to learn much about the stresses which deformed the rock. Because the types of symmetry so far observed in rocks and soils are relatively few, and contain only ∞- and 2-fold axes of symmetry, the only symmetry element observed in projection is the plane of symmetry (m). Details of fabric symmetry are shown in Table 2.7.

Table 2.7 Symmetry of fabrics in rocks

Symmetry	Planes	Example
Axial	∞	Figure 2.4a
Orthorhombic	3	Figure 2.4b
Monoclinic	1	Figure 2.4c
Triclinic	0	Figure 2.4d

Unlike the symmetry of crystallography, the determination of fabric symmetry is statistical. For this reason it is not always possible to state the symmetry precisely, since it may have aspects of more than one class.

2.5 The mechanical properties of rock material

Rock material possesses a number of physical and mechanical properties, most of which are determined by testing in the laboratory, and to a lesser extent, in the field. It is most important to be aware that values for any of these properties of the material may bear only a slight relationship, if any, to the value for that property of the rock mass. The main mechanical and physical properties of rock material are:

(a) Strength:
- (i) uniaxial compressive strength;
- (ii) shear strength;
- (iii) tensile strength;

(b) Deformation properties:
- (i) modulus of elasticity;
- (ii) Poisson's ratio;
- (iii) 'creep constants';

(c) Hydrological properties
- porosity;
- water sorption;
- moisture content;
- primary permeability;

(d) Other properties:
- (i) unit weight;
  - dry unit weight;
  - saturated unit weight;
- (ii) durability.

When testing a rock sample, it is essential that the material is intact, i.e. that it is free from defects, and also that the material is homogeneous. As it is desirable that the material tested should be typical of the mass, it may be necessary to test a large number of samples. If a linear or planar anisotropy is present in the rock material, then all of the properties listed (with the exception of porosity, water sorption, moisture content, unit weight and durability) will vary with direction.

In a mechanical sense, the strength of rock material is defined as the ability to resist stress without large scale failure. Small scale failure with the development of microfractures occurs under stresses well below the strength of the rock. Since, in rock, large scale failure only occurs beyond the elastic stress limit, this is the most commonly used parameter, especially for brittle rocks.

Uniaxial, or unconfined, compressive strength defines the failure of the rock specimen subjected to a compressive stress in one direction. The shear strength is the stress at which the rock fails in shear.

Tensile strength is the stress at which the rock fails in tension. This tensile strength is usually determined indirectly. Flexural strength is the stress at failure of a beam, supported at both ends, and loaded at the centre. As well as determining the strengths of rocks under uniaxial conditions, strengths are determined often under a confining stress: these are called triaxial tests, and attempt to simulate *in situ* conditions as well as being used to determine the shear strength of the rock.

It is sometimes assumed that rocks are perfectly elastic substances, i.e. they obey Hooke's Law:

$$\frac{\sigma}{\epsilon} = E \qquad (2.1)$$

where σ is the stress, ϵ is the strain; and E is a constant known as Young's modulus or the modulus of elasticity. In fact, very few rocks are perfectly elastic, so that E varies with stress, and the stress – strain relationship is a curve. Because of this, three values of E may be cited (Fig. 2.8):

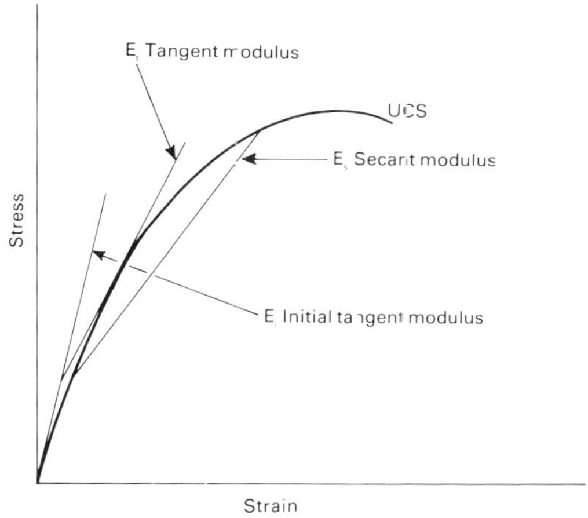

Fig. 2.8 Generalized curve showing stress–strain relationships in rock.

(i) the tangent modulus, E_t, which is the value of E at a specific stress; E_t is the gradient of the tangent to the stress–strain curve at the specific stress, usually, but not necessarily taken at 50% ultimate strength;

(ii) the secant modulus, E_t, which is the modulus over a given stress range, and is the gradient of the secant joining the two points on the stress–strain curve; and

(iii) the initial tangent modulus, a special case of E_t for $\sigma = 0$.

Another important parameter in elastic theory is Poisson's ratio, υ, defined as:

$$\upsilon = -\frac{\epsilon_x}{\epsilon_z} \qquad (2.2)$$

where ϵ_x, ϵ_z are the strains in the lateral and axial directions, respectively. As with E, the value of υ can vary with stress.

Most rocks exhibit an element of time-dependent deformation in their response to stress. Time-dependent strain, or creep, is frequently an important component of the total deformation. A rock subjected to stress suffers an instantaneous elastic strain which is completely recoverable on removal of stress. The instantaneous elastic strain is followed by creep, which will continue to occur indefinitely, but at a decreasing rate, during stress application (Fig. 2.9). When the stress is removed, only a part of the creep strain is recoverable. In general, the less brittle a rock, the greater is the creep component of the total deformation.

Porosity, moisture content, sorption and primary permeability are, to a certain extent, related. Porosity defines the pore space, or voids, in a rock as a

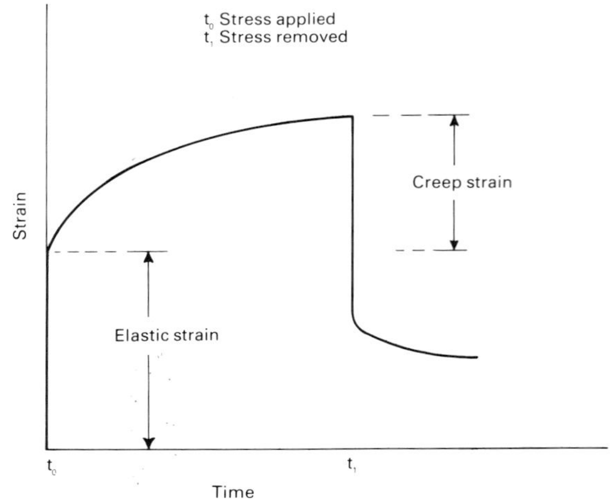

Fig. 2.9 Time-dependent deformation in rock.

ratio of the total volume of rock and voids. Sorption expresses the ability of a rock to absorb water, and primary permeability is the capacity of the *intact rock material* to transmit water. Permeability and porosity can be related as follows:

$$p = c\eta^2 \tag{2.3}$$

where p is permeability; η is the apparent porosity and c is constant. A high porosity does not necessarily mean that a rock will have high permeability and sorption since these are very much dependent on the pore fabric; size, arrangement and degree of interconnection of the pores. A shale, for example, may have porosity several times that of a sandstone, but primary permeability of the shale may be orders of magnitude less than that of sandstone. Voids in shale tend to be small and discrete, whereas those in sandstones are relatively large and are interconnected.

Moisture content of a rock is expressed as a percentage of the dry weight of the rock material. Field moisture content or saturation moisture content may each be important, depending on the circumstances. Unit weight of a rock is the weight of the material per unit volume and is usually expressed as $Mg\ m^{-3}$. Again the unit weight may refer to the material at field moisture content, or saturation moisture content. It is frequently expressed as dry unit weight, i.e. at zero moisture content. The durability of a rock material is a measure of its resistance to degradation over a relatively short term. Durability is now expressed as the slake durability index I_s, which is a measure of the degradation of the rock material under conditions of abrasion and alternate wetting and drying. Although this has tended to replace the standard sodium sulphate test of the concrete engineer, its significance and value are dubious.

2.6 Mechanical classification of rock material

Because the geological classifications of rock have such limited applications in engineering geology, many attempts have been made to develop a mechanical, or engineering, classification. No such classification, which would meet all requirements, has yet been formulated. In practical terms, the most essential requirement is a classification of the rock mass, but it is essential, as a basis for this, that the rock material be classified.

It is necessary initially, in classifying rock material, to assess its degree of homogeneity. Such an assessment is largely qualitative and subjective. In homogeneous material, the mineral constituents are so distributed that a small sample cut from any part of the material will have the same constituents in the same proportions, and will have the same properties as the material as a whole. It is also necessary to determine any mechanical anisotropy which may be present, and to specify its nature and orientation. Anisotropy is a measure of the directional properties of the rock. Statistically, the rock will be isotropic if all mineral grains have random orientation of both dimensional and crystallographic parameters. Isotropy also requires that a plane of equal dimensions intersecting the rock in any direction exposes an equal number of grains. Continuity, which is probably of more importance in a rock mass, refers to the amount and orientation of fissure space in the mass. In an ideal, so-called intact rock material, fissures are absent and the material is therefore continuous. In fact, microfractures due to weathering, as well as grain boundaries and mineral cleavages, are discontinuities. The very existence of these microdiscontinuities will influence the cohesion of the rock and also the distribution of stress in the rock material.

Classification of rock material for engineering purposes requires a statement on the following criteria:

(1) Petrography
- (a) composition;
- (b) degree of weathering;
- (c) texture;
- (d) fabric, including a statement on microdiscontinuity.

(2) Homogeneity
- (a) homogenous;
- (b) inhomogeneous.

(3) Isotropy
- (a) isotropic;
- (b) anisotropic
  - (i) number of anistropies;
  - (ii) nature and relative development of anistropies;
  - (iii) orientation of anistropies.

(4) Strength
Usually, uniaxial compressive strength only is specified, but some circumstances may also require a statement on tensile strength and on shear strength para-

meters. A standard statement on uniaxial compressive strength uses the terms:

(a) very high strength, >50 MPa;
(b) high strength, 16–50 MPa;
(c) medium strength, 5–16 MPa;
(d) low strength, 1.6–5 MPa;
(e) very low strength <1.6 MPa

(5) Elasticity
This can be expressed by the modulus of elasticity, but it is becoming increasingly common to use the *modulus ratio*, E_t/σ_{ult}, where E_t is the tangent modulus at 50% ultimate strength, and σ_{ult} is the ultimate strength. The standard descriptive terms are:

(a) high modulus ratio, >500;
(b) medium modulus ratio, 200–500;
(c) low modulus ratio, <200.

The mechanical classification of rock developed may depend on the precise purpose of the classification. The bases outlined here are intended for the laboratory study of the rock material. Deere (1968) has proposed a classification of rock material based on uniaxial compressive strength and modulus ratio (Fig. 2.10). It has been pointed out by Deere that rocks possessing an interlocking

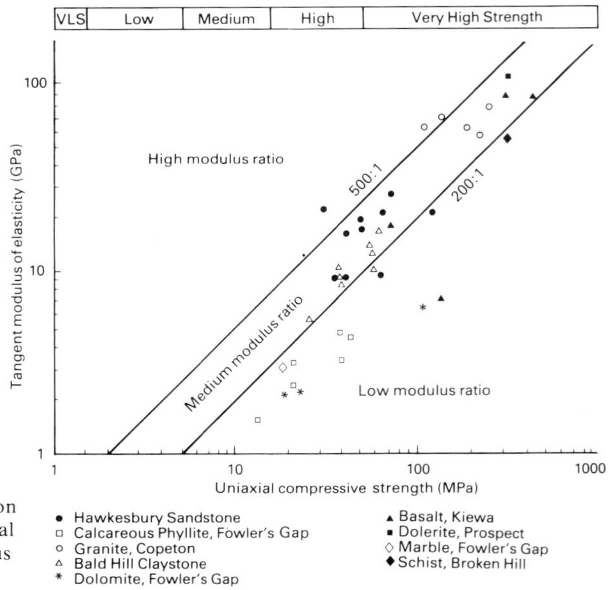

Fig 2.10 Geological classification of rock material based on uniaxial compressive strength and modulus ratio. [After Deere (1968)]

fabric, and with little or no anisotropy, fall within the field of medium modulus ratio: this includes most igneous rocks and sandstones. Rocks with a strong planar anisotropy, for example shales, phyllites and schists, may lie in either the field of low modulus ratio, if the anisotropy lies at an angle approaching 90° to the direction of loading; or they lie in the field of high modulus ratio, if the anisotropy is at a low angle to the loading axis.

This classification, although based primarily on strength and elasticity, is very useful because it is clearly sensitive to such important aspects of the rock as mineralogy, fabric and anisotropy. Complete classification of a rock material still, however, requires a brief description of lithology and fabric. For example: calcareous phyllite, high strength (HS), low modulus ratio (LM), fine grained, finely foliated.

References and further reading

Anon. (1977) The description of rock masses for engineering purposes. Report by the Geological Society Engineering Group Working Party. *Quart. Jl. Eng. Geol.* **10**, 335–88.

Blatt, H., Middleton G. & Murray R. (1972) *The Origin of Sedimentary Rocks.* Prentice Hall, New Jersey.

Deere D.U. (1968) Geological considerations. In *Rock Mechanics in Engineering Practice* (ed. K.G. Stagg and O.C. Zienkiewicz). John Wiley, New York.

Farmer I.W. (1968) *Engineering Properties of Rocks.* Spon. London.

Grim R.E. (1953) *Clay Mineralogy.* McGraw-Hill, New York

Hobbs B., Means W.D. & Williams P.F. (1976) *An Outline of Structural Geology.* John Wiley, New York.

Hyndman D.W. (1972) *Petrology of Igneous and Metamorphic Rocks.* McGraw-Hill, New York.

Mason B. & Berry L.G. (1959) *Elements of Mineralogy.* W.H. Freeman, San Francisco.

Spry A. (1969) *Metamorphic Textures.* Pergamon Press, Oxford.

Turner F.J. & Weiss L.E. (1963) *Structural Analysis of Metamorphic Tectonites.* McGraw-Hill, New York.

3

The Rock Mass

3.1 General statement

The term *rock mass* means a large volume of rock on which, or in which, some engineering work is to be carried out. If the mass is made up of a single rock material, it will be, for engineering purposes, relatively homogeneous; if it is made up of more than one material, it may be grossly inhomogeneous. Masses of igneous rock, such as granite, are often homogeneous, although volcanic rock masses may not be. On the other hand, masses of sedimentary and metamorphic rocks are usually inhomogeneous in the extreme. No rock mass is truly continuous: it will be broken by cooling and shrinkage joints and fractures, bedding separation surfaces, tectonic joints and faults, and minor igneous intrusions such as hydrothermal veins, dykes and sills. In addition, the near-surface segment may be weathered. The intensity and nature of the discontinuities depends on the origin, nature, homogeneity and geological history of the rock mass. Discontinuities in an initially homogeneous rock mass not only contribute to a destruction of continuity, but in a reduction in homogeneity since it is along fractures that dykes and sills are intruded, and that weathering is most active. Faulting produces a cataclastic rock which differs from the parent rock, in that it is completely crushed, and may have different mineral assemblage; while faults and unconformities may bring into contact rocks which are completely different, in all respects, from each other.

3.2 Homogeneity and isotropy of the rock mass

From a strictly geological point of view, even the smallest sample of rock lacks true homogeneity and true isotropy. If this is the case, then neither homogeneity nor isotropy could be expected in a large mass of rock. While this is true, the engineering geologist and engineer are concerned with inhomogeneity and anisotropy sufficiently developed to exert some influence on the engineering properties of the rock mass. In these terms, homogeneity and isotropy are more likely to occur in igneous rock masses and inhomogeneity and anisotropy in sedimentary and metamorphic masses.

3.2.1 IGNEOUS ROCKS

If chilled border phases, such as are developed about the margins of intrusive masses, are excluded, most plutonic rock masses can be homogeneous with respect to both composition and texture on a scale of significance for the engineer. It is true that if the mass is closely examined inhomogeneities in the form of xenoliths and mineral segregations will be found, but these will be of little, and usually of no, significance in the mechanical behaviour of the rock mass. It is reasonable to assume that, in the fresh condition, plutonic masses of granite, syenite, diorite, gabbro and related rocks are homogeneous, although some plutonic masses may be layered, e.g. gabbro. It is also a reasonable assumption that, unless flow foliation is present, these rocks are isotropic at the mass scale. Some plutonic masses have undergone such an intrusive history that flow structures are sufficiently strong to impose a quite definite planar anisotropy on the rock mass.

Most masses of volcanic rock are made up of a series of individual flows, and, while the differences between individual flows may not be great, it is common to find between each successive flow such features as ash beds, fossil soil horizons, and the broken fragmental rock characteristic of lava flow surfaces. At least until detailed studies are completed, it is sound practice to assume that volcanic masses are inhomogeneous. This is not to say that, within some very thick individual flows, homogeneity will not be found.

Since flow is an essential process in the formation of volcanic rocks, flow structures are often strongly developed throughout the mass. The mass, then, can be expected to possess some degree of linear or planar anisotropy. In some cases this is sufficiently strong to have mechanical significance; in others it is weak, and of little or no importance.

3.2.2 SEDIMENTARY AND METAMORPHIC ROCKS

Sedimentary rocks, and metamorphic rocks derived from them, are layered sequences, each layer in the sequence differing in composition, texture, and fabric from those above and below it. Some sedimentary layers are graded beds, with coarse material at the base grading up to fine material at the top, while beds of shale contain internal laminations of differing texture. Sedimentary layers within themselves may, then, be inhomogeneous; the whole mass is strongly inhomogeneous. Similarly, metamorphic rocks derived from sedimentary rocks will be inhomogeneous. The possible exception is to be found in some masses of marble, which are relatively homogeneous, these having been derived from a homogeneous limestone mass. Metamorphic rocks derived from a homogeneous igneous mass tend to inherit the homogeneity of the parent rock, but they may develop a planar structure due to deformation, e.g. slates.

If, during metamorphism, metamorphic differentiation of a type which leads to lithological layering takes place, then the degree of inhomogeneity may be increased. This also tends to increase both the intensity and complexity of the planar anisotropy which exists in most sediments. The development of cleavages, schistosity and lithological layering may increase the complexity since these metamorphic structural elements may be imposed with orientations different to that of the original bedding, and sometimes different to each other. Some high grade granulites and amphibolites may be both homogeneous and isotropic, and while some marbles may be homogeneous, their isotropy is apparent rather than real. Careful study usually reveals quite strong mechanical anisotropy, due to deformation of these rocks during their formation.

3.3 Continuity

It has already been stated that perfect continuity is not to be found in any rock mass. In sedimentary rock masses the discontinuities are bedding separation surfaces, joints and faults; in metamorphic rocks, inherited bedding surfaces, joints and faults; in igneous rocks, joints and faults, with flow separation surfaces in volcanic masses. All rock masses, irrespective of origin, may contain unconformities, and dykes and sills as small intrusions. The boundary of a plutonic mass constitutes a major discontinuity. Discontinuities such as bedding, joints and faults have a definite pattern, although the pattern may vary throughout the mass. Joint patterns, especially, are more intense and more complex near the surface, where weathering is active, than they are at depth. The effect of discontinuities is to break a monolithic mass into a series of layers and blocks. The mechanical importance of any system of fractures will depend on their nature, spacing, orientation and condition. Some writers refer to discontinuities in a rock mass as *defects*; this is an apt term, since their presence effects a mechanical deterioration of the mass in relation to the material. Moreover, the occurrence of discontinuities leads to variations in the stress distribution in the mass, and may lead to stress concentrations of critical magnitude in the rock about excavations.

3.3.1 BEDDING SEPARATION SURFACES

When the sediments are first deposited, bedding separation surfaces are usually planar and horizontal. They form the boundaries between layers, or beds, of sediment which differ from each other in texture and composition. Each

Fig. 3.1 Sedimentary beds folded into an anticline, Cape Liptrap, Australia.

individual bed probably represents a single depositional event. With tectonic deformation of the sedimentary mass, the surfaces become tilted and folded (Figs 3.1, 3.2) and sometimes show a high degree of complexity. At depth,

Fig. 3.2 Sedimentary beds folded into a syncline, Cape Liptrap, Australia.

bedding planes are usually tightly closed, and become open only with weathering, and with stress release due to erosional unloading. Spacing varies widely from a few millimetres to many metres. Under some circumstances, sandstones may contain two sets of bedding surfaces; those separating the beds, and an internal set, usually quite steeply inclined, which are due to current action during deposition. This internal bedding is referred to as false bedding, current bedding, or cross bedding (Fig. 3.3).

Fig. 3.3 Current cross bedding in sandstone, Break O'Day, New Zealand.

Bedding separation surfaces, like other fractures in a rock mass, may contain infillings, and are certainly a locus for weathering. They are rarely smooth surfaces, and sometimes are quite wavy due to the existence of ripple marks, or they may be quite irregular due to basal deformation of the beds (Fig. 3.4).

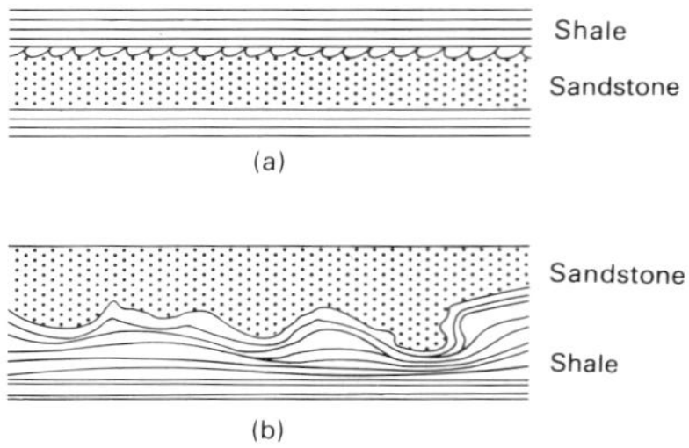

Fig. 3.4 (a) Ripple marks on upper surface of sandstone bed giving uneven base to overlying shale. (b) Load structures in base of sandstone bed leading to irregular separation surface and distortion of layering in underlying shale.

3.3.2 JOINTS

Joints are imposed on rock masses in a variety of ways. In an igneous rock, they may be due to contraction with cooling; in a sedimentary rock, to shrinkage with drying; and in all rocks, to tectonic stresses (Fig. 3.5). Weathering and the

Fig. 3.5 Jointing in sedimentary beds, Cape Liptrap, Australia. Note the variation in joint patterns and spacing in the different beds.

associated erosional unloading produce an open joint structure, known as sheeting, which is most prominent in massive crystalline rocks. Irrespective of their origin, joints are defined as fractures, in a rock mass, along which little or no displacement has occurred. Again irrespective of their origin, joints normally occur as systems of relatively regular sets. Below the zone of weathering, they are usually tightly closed, opening only as a result of weathering. Joint spacings may be clean, or they may be infilled with secondary mineral materials, and the surfaces may have a coating of clay or other minerals. These secondary mineral deposits may be due to movement, to hydrothermal alteration, or to weathering. The more open joints at, and near, the surface, may carry an infilling of debris washed in from the surface.

The mechanical properties, and importance, of joints depends on their nature, tightness, orientation, pattern, spacing and on whether or not infillings are present, and if so, on the nature of the infilling. Joint openings provide channels for the circulation of groundwater through rock masses, the material of which is virtually impermeable. It is the joints which give the rock mass its so-called secondary permeability.

In engineering geological studies, recognition of the origin of a given set or system of joints is important other than for purely geological reasons. This is especially true when joint systems are to be analysed for purposes of residual stress analysis. It must be accepted that it is often difficult, and sometimes impossible, to state with certainty the origin of some joint sets in a rock mass.

3.3.3 FAULTS

The geologist's definition of a fault as being a fracture, along which significant displacement has occurred, is an oversimplification. Fault movements are complex and, in most cases, the resulting structures reflect this complexity. Most faults of any magnitude do not consist of a single fracture, but of an intricate system of fractures, intense jointing of the wall rocks, and a crush zone, or shear zone in which cataclasis has produced a rock with inferior mechanical properties to the wall rock (Figs 3.6, 3.7). Cataclasis is a feature of thrust and wrench faults

Fig. 3.6 A low angle thrust fault in the No. 4 Tail Race Tunnel, Kiewa, Australia. Crushed Palaeozoic gneiss has been thrust over Quaternary alluvium.

especially; the rocks produced may be strongly foliated flow mylonites, fault breccia, fault conglomerate, or gouge. The thickness of the zone of cataclasis, or crushing, can range from a few centimetres to over a kilometre. Similarly, displacement along a fault can be a few centimetres or it may be hundreds of kilometres.

Faults have significance to engineering geology in a number of ways:

(1) They constitute major discontinuities in a rock mass.

(2) Entirely different rocks may be brought into contact, destroying the homogeneity of the rock mass.

(3) They can affect, to a marked degree, the stress distribution in a rock mass.

(4) The zones of cataclasis comprise materials with poor mechanical properties.

(5) Faults may be active and capable of subjecting the rock mass to dynamic stresses.

Fig. 3.7 Flat thrust fault, with crush zone of mylonite. Note foliation developed in mylonite, and the jointing of the wall rock, Tawonga, Victoria, Australia.

All faults are due to tectonic activity of some type, and, while the effects of tectonic stresses may appear to be concentrated in the crush zone, the stresses cause jointing to be developed in the wall rocks which is not only genetically, but also geometrically, related to the fault. Adjacent to a crush zone, jointing is more intense, with a pattern different to, and more complex than, that of the main mass.

Crush zones may constitute groundwater aquifers in an otherwise impermeable mass. This is generally the case if the fault rock is a breccia or congolomerate. Zones of gouge and mylonite are usually of low permeability and may act as aquicludes or aquifer boundaries.

3.3.4 DYKES AND SILLS

Dykes and sills are intrusive sheets of igneous rock which have been emplaced along joints or bedding surfaces, or similar types of opening, at a time when the rock mass is under tensile stress. These structures, which are considered to be discontinuities, vary in length and thickness. Some are only a few centimetres thick, and persist for only a few metres; others are hundreds of metres thick and persist for hundreds of kilometres. The dolerite sills of Tasmania and the Whin Sill of Britain are examples of the larger structures. The large dykes or sills constitute a rock mass in themselves, and possess all the characteristics of a rock mass. It is the smaller structures, with thickness of the order of centimetres or metres, which are commonly regarded as discontinuities.

Dykes and sills frequently occur as 'swarms' and have a quite definite pattern, often related to the fracture pattern of the rock mass. For example, the swarm of lamprophyre dykes of south-eastern Australia is very closely related, geometrically, to the pattern of faults and joints in the rock masses. While these structures constitute a discontinuity in the rock mass, their existence may be beneficial in as much as they may 'heal' fractures in the rock. More often, however, they create geomechanical and engineering problems: e.g. Beavis (1949, 1962) records harmful effects of dykes at a dam site, and in underground hydroelectric power stations.

3.3.5 UNCONFORMITIES AND INTRUSIVE CONTACTS

Marked time breaks in the geological history of an area, when the rock mass is subjected to erosion, occur from time to time. When the rock mass is then subjected to depression, with the deposition of another sequence of rock, the old erosion surface constitutes a major discontinuity, known as an unconformity. Unconformities are normally of quite irregular topographic configuration, and bring into contact sharply contrasting rock masses. For example, two unconformities north of Broken Hill, Australia, have been mapped: on one, highly deformed pre-Cambrian crystalline schists are overlain by gently dipping Devonian sandstones and shales; the second forms the upper boundary of the Devonian sediments, above which are horizontal unconsolidated Tertiary to Quaternary sands, clays and gravels.

The boundaries of intrusive igneous bodies are usually irregular, and they may be steep or flat. Although it is sometimes the case that these boundaries are modified by faulting, the more usual situation is that the boundary is quite sound, in engineering terms, since the magma almost certainly has reacted with the rock which has been intruded. Although these contacts, as they are usually called, mark a boundary between two dissimilar rocks, they are not otherwise, normally, of engineering significance. It is sometimes the case that border zone reaction produces a contaminated material within the margin of the igneous mass which has a lower resistance to weathering than either the main mass of igneous rock, or the thermally metamorphosed country rock, which forms an aureole about the intrusion.

3.4 Physical and mechanical characteristics of discontinuities

Fractures, and other discontinuities in a rock mass, possess a number of physical and mechanical characteristics, some of which can be measured and which influence the behaviour of the rock mass when placed under stress. These characteristics include the number of sets; the location and orientation of the fractures, and their spacing in both two and three dimensions; the nature of the openings and of surfaces and infillings; and the persistence or continuity of the fractures.

3.4.1 NUMBER OF DISCONTINUITIES

A set of fractures consists of all of those which have similar orientation. A system is a uniform group of two or more sets. Some fractures and other discontinuities, such as joints and bedding planes, occur as sets, and can be analysed statistically. Others, such as faults, dykes and unconformities are unique as a general rule, and have to be considered individually. Notwithstanding this, faults and dykes can, and not infrequently do, occur as sets and systems.

The existence of discontinuities in a rock mass will have the effect of changing its properties. The degree to which this will occur depends on the number of sets, or of unique discontinuities, present in the mass. The most important aspect is the ability of the rock to deform without involving the fracture of the rock material; this is highly dependent on the number, and orientation, of discontinuities present in the mass.

3.4.2 LOCATION AND ORIENTATION

Ideally, each individual discontinuity should be located and its orientation measured, and both location and orientation recorded on a map. This is desirable since a single discontinuity may be of critical importance. In practice, however, especially in the case of joints and bedding planes, such a procedure is impossible, except on the very largest scale and most detailed maps. It is more usual to note the location, or small area, in which the observations were made, and to state the nature and orientation of the sets for that location.

The orientation of discontinuities is one of their more important properties, since it is often this aspect which will control the stability of an excavation in the rock mass. For example, failure by sliding is more likely to occur on a discontinuity which dips towards an excavation than on one which dips away from the excavation (Fig. 3.8). It is well known that the stability of large

Fig. 3.8 Effect of dip of discontinuities on stability of a cut slope. Planes marked 'a' are surfaces of potential sliding failure.

underground excavations is often controlled by the orientation of the discontinuities in relation to the configuration of the excavation.

3.4.3 SPACING OF DISCONTINUITIES

Fracture spacing is determined by counting the number of fractures which cut a selected length of traverse line. A single traverse line in any area is usually quite inadequate (Fig. 3.9); at least two traverse lines are required at right angles to each other. Preferably traverse lines should be set out at right angles to each of the sets present. Even so, the result obtained may depend on the actual location of the traverse lines.

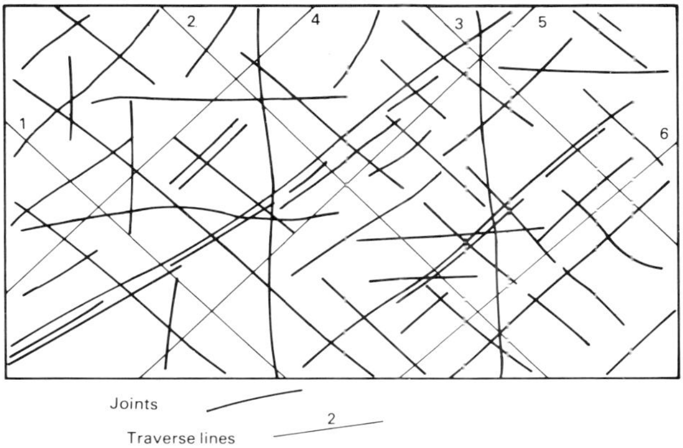

Fig. 3.9 Suggested traverse pattern for fracture spacing surveys in the field.

Spacing can be expressed in terms of the average distance between the fractures, or, if more detail is required, a large scale fracture map may be prepared. The spatial distribution intensity map (Fig. 3.10) is probably the most

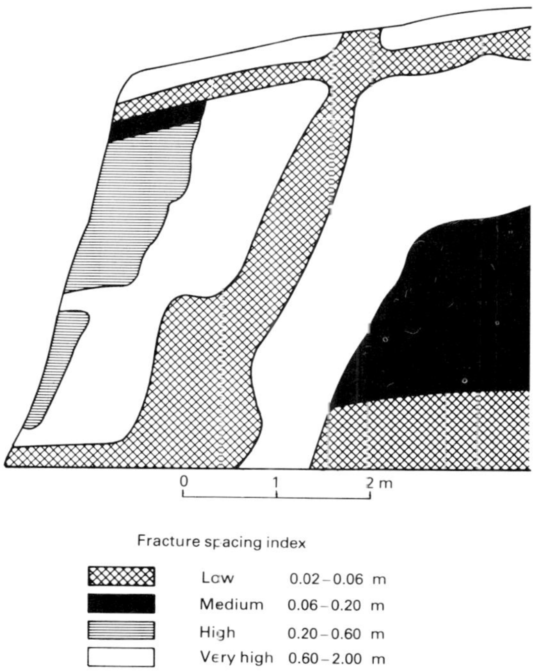

Fig. 3.10 Fracture spacing intensity map for cutting in Hawkesbury Sandstone, Southern Tollway, Sydney, Australia.

effective method of presentation; moreover, it may be of more value to the engineer than a very detailed fracture map. It is always essential to consider the exposure level when analysing fracture systems.

Fracture spacing is usually expressed as a fracture spacing index I_f (Table 3.1). This considers specifically the spacing of fractures in two dimensions.

Table 3.1 Fracture spacing index. The simple logarithmic fracture spacing index is defined as the average size of cored or outcrop material.

Class		Fracture Spacing Index I_f (m)
Extremely high	EH	>2
Very high	VH	0.6–2
High	H	0.2–0.6
Medium	M	0.06–0.2
Low	L	0.02–0.06
Very low	VL	0.006–0.02
Extremely low	EL	<0.006

Another index in use is the area intensity index (Table 3.2), in which the area of fracture surface per unit volume of rock, and the average size of the intact block, are measured. Of these two indices, the former is certainly the simpler to measure both in outcrop and core; the area intensity index requires exceptionally good and well-placed excavations or very good natural exposures.

Table 3.2 Area intensity fracture index.

Class		Area of fracture surfaces per unit volume (m^2/m^3)	Average volume of intact blocks
Very low	VL	<3	>1 m^3
Low	L	3–10	0.027–1 m^3
Moderate	M	10–30	0.001–0.027 m^3
High	H	30–100	27–10^3 cm^3
Very high	VH	100–300	1–27 cm^3
Complete	C	≥300	≤1 cm^3

It is found, in practice, that the area intensity index is easier to apply to drill core than to field exposures, and that the fracture spacing index is more useful. It is the fracture spacing index which is coming to be accepted as the standard.

3.4.4 OPENNESS AND INFILLING OF DISCONTINUITIES

Below the zone of weathering, all discontinuities of the fracture type are usually tight due to the state of stress in the rock mass, including confining pressure. If a redistribution of stress occurs due to excavation, or with natural weathering and erosional unloading, then the fractures tend to open, and to become infilled with material produced by weathering and deposited by circulating groundwater. In some cases, infillings are present even before these processes operate: cataclastic material on faults and on bedding plane shears, and fine crushed material, usually highly polished, on shear type joints. It is usual for fracture openings to increase in width up through the weathered zone, reaching a maximum at the surface. For those discontinuities such as faults, dykes and sills, in which the separated surfaces are infilled with igneous or cataclastic material, thickness does not bear any relationship to weathering or proximity to the ground surface. This is not to say that thickness is constant: this can vary between wide limits for a particular structure, and change can occur very suddenly.

A standard set of terms is in use to describe the openings of fracture discontinuities and thickness of faults and dykes (Table 3.3).

Table 3.3 Openness and thickness of discontinuities.

Description	Width or thickness (mm)
Wide	>200
Moderately wide	60–200
Moderately narrow	20–60
Narrow	6–20
Very narrow	2–6
Extremely narrow	>0–2
Tight	0

The significance of the openness of fractures lies in the increased secondary permeability imparted to a rock mass as width increases, and in the decrease in stability with increasing width.

The infilling in faults may be gouge, breccia or fault conglomerate. In the case of bedding plane shears, the material is usually a very fine gouge. In joints, and along normal bedding planes, weathered rock is characteristic, but other mineral matter may also be present, either washed in by circulating waters, or deposited from solution from these waters. Some shear faults carry a film of highly polished and slickensided gouge. The resistance to sliding along a discontinuity is a function of the openness and particularly of the infilling, both its composition and its thickness. It is essential that the properties of the infilling material be determined.

Two special cases of discontinuity must be considered. The first of these is the solution cavity which occurs in limestones, marbles, and related rocks. These may be quite small and be concentrated along bedding planes or other fractures, or they may be of great size and quite irregular in shape. Although many exist as open caves, others may be more or less infilled with material washed in from the surface, and by fallen blocks of rock. The second case is that of sheeting, which is an erosional feature most commonly found in massive crystalline rock masses. Sheeting occurs as sets of parallel discontinuities, sub-parallel to the natural surface, with spacing between the individuals decreasing with depth (Fig. 3.11). These structures carry varying thickness of completely weathered rock, and thickness of the structure is very sensitive to loading. Beavis (1962) recorded increases in thickness by up to 50% over a 90 day period, when excavation was

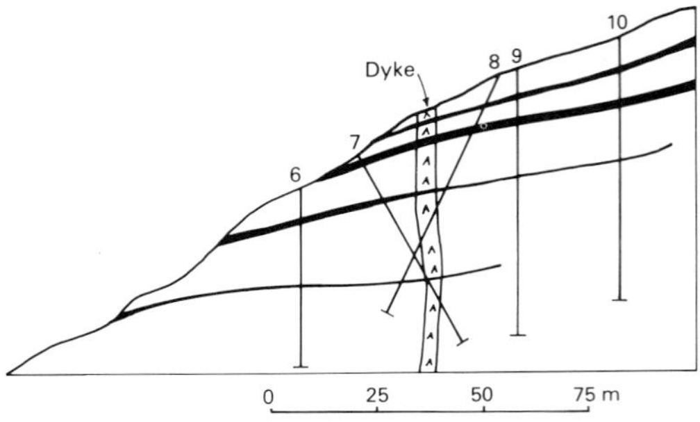

Fig. 3.11 Cross section of part of the west abutment, Clover Dam, Kiewa, Australia, showing sheeting as observed in outcrop and inferred from bores. The two lower structures caused instability during foundation construction.

carried out over sheeting sets. Sheeting often controls the stability of slopes, and is of particular significance in the abutments of dams.

3.4.5 PERSISTENCE OF DISCONTINUITIES

Some discontinuities such as faults, dykes, unconformities and bedding planes can be traced along their strike and down dip for considerable distances. Joints may be of quite limited extent; equally, they may also persist for considerable distances. The persistence of a discontinuity gives some measure of the proportion of rock material which would have to be sheared during failure along a surface. Persistence is an important property, but one which is difficult to measure and difficult to quantify. Every field geologist knows the difficulty of tracing a joint or bedding plane beyond the limits of an outcrop.

It is in inhomogeneous rock masses that joints tend to be least persistent. The pattern of jointing is sensitive to compositional and textural characteristics of the rock material, so that, in response to any stress, the joint pattern of the rocks will vary from bed to bed, and it is only rarely that a joint crosses a bedding plane (Fig. 3.5). In graded beds, joints may even be curved, while the fine defects such as cleavage are not only curved across graded beds, but are refracted at bedding planes.

3.4.6 NATURE AND CONDITION OF DISCONTINUITY SURFACES

The condition of the rock forming the surface of a discontinuity can have a profound effect on the shear strength of the discontinuity. While the rock mass as a whole may show little or no weathering, this can occur to quite an advanced degree along, and adjacent to, a discontinuity. The actual surface can be

Table 3.4 Categories of roughness of discontinuity surfaces.

Category	Condition of surface
1	Polished
2	Slickensided
3	Smooth
4	Rough
5	Defined ridges
6	Small steps
7	Very rough

described generally as wavy, rough, or smooth, although seven categories are now recognized (Table 3.4).

The difference between waviness and roughness is largely one of degree, although they also differ in their effects on the shear strength of the discontinuity or the characteristics of the discontinuity during shearing. Waviness refers to those first-order asperities which appear as undulations of the plane and which would be unlikely to shear off during movement. Roughness refers to second-order asperities which because of their size would be sheared off during movement (Fig. 3.12). Methods of measurement of wavelength and amplitude are numerous, but Fecker and Rengers (1971), and Fecker (1978) have described a method which is most effective.

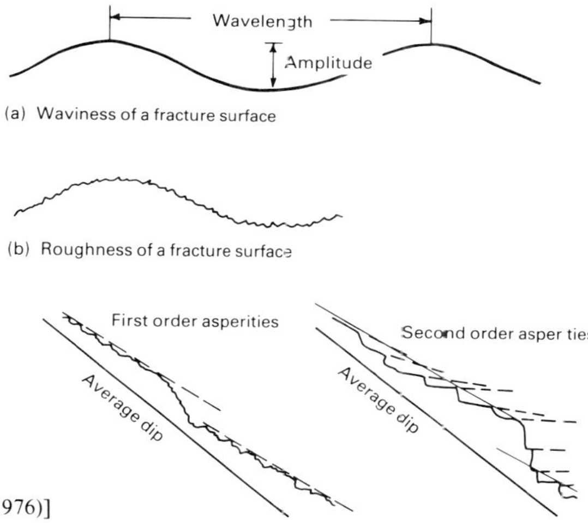

Fig. 3.12 Asperities on a fracture surface. [After Attewell and Farmer (1976)]

The orientation of the asperities on the surface ranks equally in importance with their magnitude, since the effect on shearing resistance along the fracture will depend on the orientation of the asperity in relation to direction of movement. For example, a considerably lower resistance to shearing would exist in a direction parallel to the asperities than in a direction normal to them.

3.5 Classification of discontinuities

Discontinuities in a rock mass may be classified on a genetic, geological basis (Table 3.5), which is essential, but only partly satisfactory, for engineering purposes, or they may be classified on the basis of mechanical characteristics (Table 3.6).

The importance of the geological genetic classification lies in the necessity to specify the nature of the discontinuity in describing a rock mass, just as it is necessary to specify the material constituting the mass. It is essential, for example, to distinguish between a joint and a fault, and between a joint and a bedding plane. Such specification alone, however, is inadequate, and mechanical and physical characteristics must be measured and described.

Table 3.5 Genetic classification of discontinuities.

Rock group	Syngenetic	Epigenetic
Igneous	Cooling joints Flow surfaces Plutonic contacts Dykes, sills, veins	Faults Tectonic joints Sheeting
Sedimentary	Bedding planes Internal laminations Cross bedding Shrinkage joints Unconformities	Faults Tectonic joints Cleavages Bedding plane shears Sheeting
Metamorphic	Foliations Metamorphic layering Joints	Faults Tectonic joints Sheeting

3.6 Analysis of fracture systems

Analysis of fracture systems is, in the first instance, a geometrical analysis; the validity of this analysis depends on firstly, the certain recognition of the origin of the fractures; and secondly, on the system being homogeneous, i.e. that the pattern, at least statistically, is uniform for the mass under investigation. If it is not, then the mass must be divided into units, or sub-areas, for each of which the fracture pattern is uniform. If the rock mass itself is homogeneous, it is likely that the fracture system developed in it will be homogeneous. If the mass is inhomogeneous, the fracture pattern certainly will be inhomogeneous.

3.6.1 GEOMETRIC ANALYSIS OF FRACTURE SYSTEMS

A set consists of all of those fractures which have (approximately) the same orientation. A system comprises a regular combination of two or more sets. Since, even within a set, orientations show some variation, statistical analysis of field data is necessary in most cases to establish the average orientation of a set, and to determine the sets which constitute a system. Several methods of analysis are used:

(i) the fracture rose, in which the strikes only are considered (Fig. 3.13). This

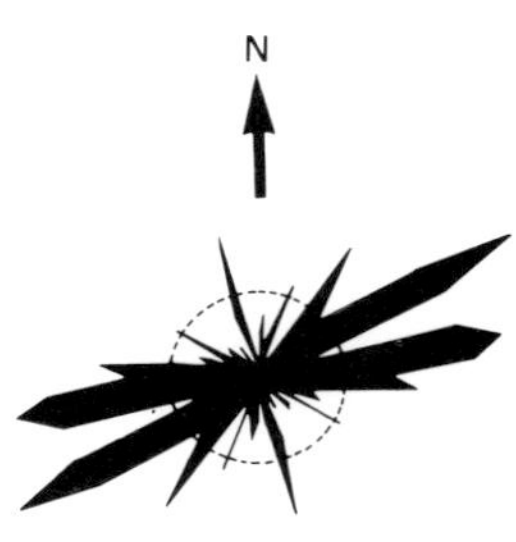

Fig. 3.13 Joint rose for a Devonian sandstone near Broken Hill, Australia.

technique takes no account of dip, nor can it include horizontal fractures. Since, in determining the mechanical characteristics of a rock mass, horizontal

Table 3.6 Mechanical classification of discontinuities. (After Fecker 1978.)

Characteristic	Faults	Shear joints	Tension joints	Bedding planes	Dykes and sills	Igneous	Sheeting
Occurrence	Generally unique	Sets	Sets	Sets	Generally unique	Unique	Sets
Continuity	High	Medium	Low	High	High	High	High
Length	Extremely long	Long to short	Medium to short	Very long	Very long to extremely long	Very long	Medium
Condition	Tight, smooth to polished surfaces	Tight, smooth to polished	Open, rough surface	Tight, rough surface	Tight, smooth to rough surfaces	Often gradational Tight	Open
Infilling	Cataclastic material	Gouge			Igneous rock		Weathered rock
Movement	Slickensides	Slickensides		May be rarely slickensided	May be slickensided		
Fracture angle 2α	Varies depending on nature of fault						
Friction angle		Peak 30–50° Residual 20–40°	Peak 40–50° Residual 30–45°				
Others				May be horizontal, tilted, vertical or folded	May be of such dimensions as to constitute a rock mass	Sometimes faulted in whole or in part	

fractures are at least as equally important as dipping fractures, their exclusion from the analysis has serious consequences. Similarly, two sets may exist which have the same strike, but quite different dips: the rose analysis fails to distinguish between the two sets. This technique probably has its greatest application in preliminary studies of aerial photographs.

(ii) the equal area analysis, in which the poles to the fractures are plotted on an equal area net, and concentrations are brought out by counting and contouring (Fig. 3.14). This method takes account of both strike and dip, and

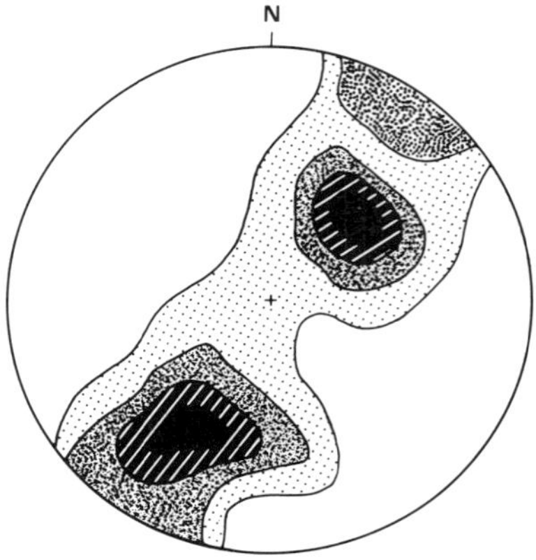

Fig. 3.14 Equal area diagram of poles to joints (75 joints) in a metaquartzite near Broken Hill, Australia. Contours 1–2–3–4%.

horizontal fractures are included in the analysis. Such an analysis is to be regarded as essential in the geomechanical study of a rock mass: it permits the definition of sets and systems, and forms the basis for kinematic analyses.

3.6.2 STRESS ANALYSIS

As well as serving to define a fracture pattern, and to isolate sets, the equal area method is used for the qualitative analysis of stresses in a rock mass. The ideal fracture system resulting from a system of stresses is shown in Fig. 3.15. It is this which is the basis of the stress analysis. The fractures intersect in σ_2, the intermediate principal stress; σ_1, the maximum principal stress is the acute bisector of the shear fractures, normal to σ_2 and lying in the $\sigma_1\sigma_2$ plane; and σ_3 is normal to the $\sigma_1\sigma_2$ plane.

The fracture system having been determined by equal area analysis, the sets which make up the system are extracted from the diagram and plotted in cyclographic projection on a stereographic net (Fig. 3.16). The intersection of the lines defines σ_2. σ_1 is then determined as the acute bisector, normal to σ_2, of the shear fractures. In this way, the orientation of the stress field responsible for the fracture system can be established. The validity of this type of analysis is often extremely doubtful. It has to be assumed that all of the fractures analysed are due to one tectonic incident; it also has to be assumed that the residual stress field has the same orientation as that responsible for the fractures. Neither assumption is necessarily valid. Nonetheless, provided the restrictions imposed by these

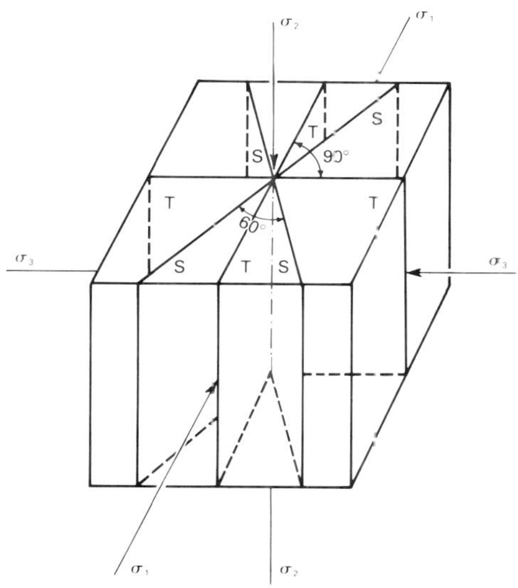

Fig. 3.15 Ideal system of tectonic joints: $\sigma_1 > \sigma_2 > \sigma_3$.

assumptions are borne in mind, the analyses can give useful results for preliminary stress studies. The method can be extended to faults and associated syngenetic joints (Fig. 3.16b): in this case not all of the essential data are necessarily always available.

3.7 Permeability of the rock mass

The primary permeability of rock refers to the flow of water through the intact rock material under the influence of hydraulic pressure. For practical purposes, many rock materials are virtually impervious. However, a rock mass, composed of such impervious material, may, as a mass, have quite high permeability as a result of fractures, other discontinuities, and solution cavities. This is known as secondary permeability. A number of methods for determining secondary permeability have been developed, ranging from such simple tests as the time taken for the water level in a bore to fall 1 m, to quite refined instrumental measurements. The most generally used test is the so-called Lugeon, or packer, test.

The Lugeon test is carried out in a bore hole in which a packer, or a double packer, is used to seal off a section of the rock mass. Water is pumped under pressure into the test section, and, after allowing for saturation of the rock, the loss of water over a specified period is determined. By definition, the test should be carried out over a test section 1 m long under a pressure of 1 MPa. A loss of 1 l min^{-1} represents 1 Lugeon (approximately a coefficient of permeability of 10^{-4} cm s^{-1}). Ideally the test should be carried out in a hole of NX size, but experience has shown that the size of the hole is not critical.

Secondary permeability depends on the spacing, continuity, infilling, openness, and degree of interconnection of the discontinuities in the rock mass. For this reason, experienced workers can, from a knowledge of the fracture

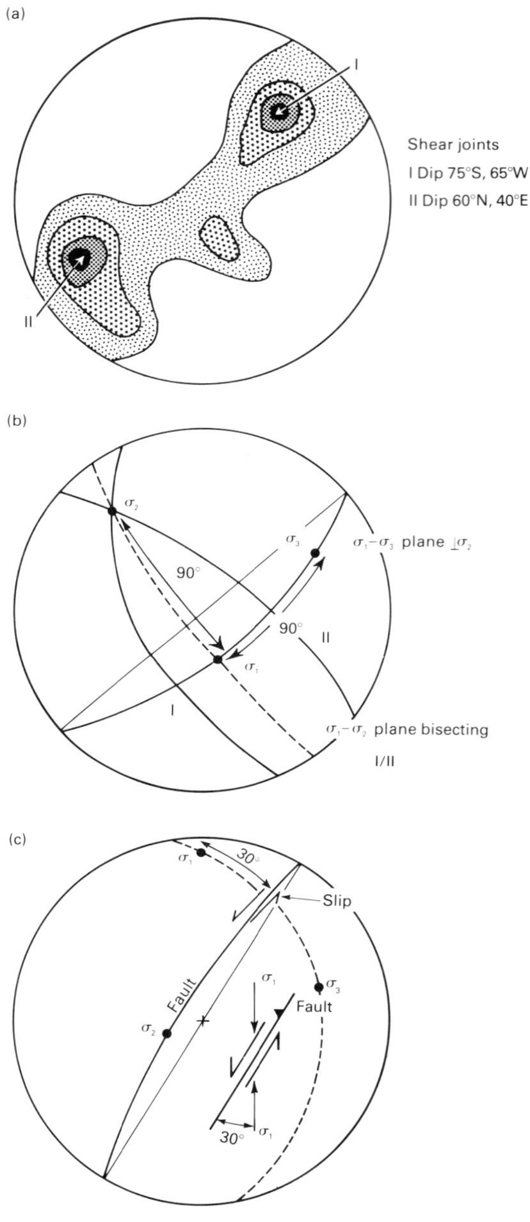

Fig. 3.16 Stress analysis from fracture data: (a) Poles to joints in a granite at Yackandandah, Australia. (b) Analysis of data from Fig. 3.16a to determine orientation of stress field. (c) Analysis of stress about a fault whose slip direction is known. The orientation of σ_1 is assumed.

pattern, make quite reliable estimates of relative permeability of the rock mass. Despite this, tests should normally be carried out. Secondary permeability can undergo quite major changes due to engineering works, especially due to excavations at depth where, due to stress redistribution in the mass about excavations, fractures which were quite tight may open to a greater or lesser degree.

3.8 Seismic velocity in the rock mass

Seismic velocity refers to the velocity of propagation of pressure waves through a rock mass. Seismic velocity is not an independent variable, but is a function of a multitude of rock properties which includes density, porosity, composition, cementation and degree of fracturing. In the case of fractured rock, and of weathered rock, it is often possible to correlate variations in degree of fracturing, or degree of weathering, with seismic velocity. For example, within the Sydney Basin, Australia, completely weathered sandstone has a seismic velocity of 860 m s^{-1}; moderately weathered sandstone, 2290 m s^{-1}; slightly weathered sandstone, 3100 m s^{-1}, and fresh sandstone, 3310 m s^{-1}.

The effect of discontinuities in a rock mass can be studied by comparing the velocity of the compressional wave in the rock mass, with the sonic velocity of intact rock measured on a core in the laboratory. The velocity ratio V_F / V_L, where V_F is the velocity measured in the field, and V_L that measured in the laboratory, should be close to unity for a fresh rock mass with only a few, tight discontinuities. As the degree of fracturing increases, V_F / V_L will reduce to values less than 1. A claystone from Sydney, with fracture spacing index, I_f=0.1 m gave V_F / V_L =0.975; the rock with I_f = 0.02 m had V_F / V_L = 0.295. Velocity ratio is also influenced by the degree of weathering. Hawkesbury Sandstone tests gave the following values for V_F / V_L: F: 0.962; SW: 0.787; HW: 0.352.

Deere (1968) has suggested a close correlation between velocity ratio and rock quality designation, but data are not adequate to establish, with certainty, that such a correlation always exists.

3.9 Rock Quality Designation (RQD)

It has been established that, provided good drilling techniques are followed, a relationship exists between the amount of core recovered and the fracture spacing, weathering, or other softening of the rock. Since fracture spacing and weathering are two of the most significant features determining the engineering quality of a rock mass, core recovery has been seen as a measure of rock mass quality. Experience has shown that core recovery, in itself, is not always a reliable indicator of quality: however, a *modified* core recovery, in which only those pieces of core of length 10 cm or more are measured, has been found to be a reliable measure of quality. This quality measure is the rock quality designation, RQD, and is the total length of core comprising lengths of 10 cm or more, expressed as a percentage of the length drilled. It is obvious that, if core recovery is poor, the RQD will be reduced if drilling techniques are unsatisfactory, even if the rock is of good quality.

Based on RQD, standard terms have been adopted to describe the quality of the rock mass (Table 3.7).

These terms must not be interpreted literally for every circumstance, since the term quality itself is flexible. Good quality rock for a building foundation may be fair, or even poor, for a pressure tunnel. Nor does the use of RQD exonerate the geologist from attempting to solve the problem of low core recovery: core not recovered is often of greater significance to the engineer than the core that has been recovered and is available for inspection.

Table 3.7 Rock quality and RQD

RQD (%)	Description of rock quality
0–25	Very poor
25–50	Poor
50–75	Fair
75–90	Good
90–100	Excellent

Some workers have established a good correlation between RQD and fracture spacing, which would be expected (Fig. 3.17a). However, the correlation is not always good (Fig. 3.17b) since fractures vary in condition, while the drilling

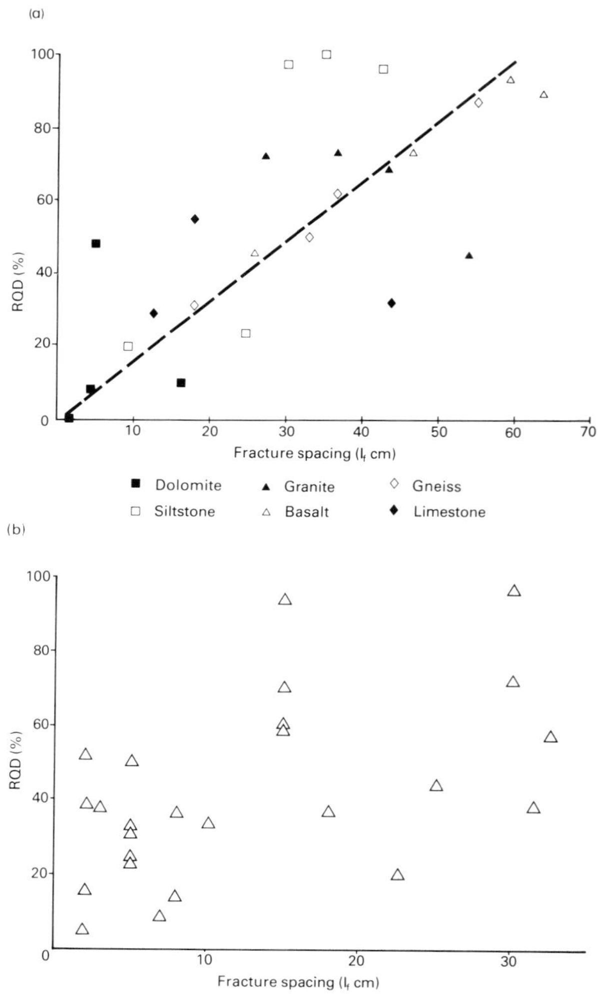

Fig. 3.17 (a) Relation between RQD and fracture spacing index for six Australian rocks. A correlation is suggested, but data are inadequate for detailed statistical analysis. (b) Relation between RQD and fracture spacing index for calcareous phyllite. Fowlers Gap, Australia. Correlation here is very weak. Data were obtained from 90 m of NX core.

process may effect some influence. Core will break on many fractures, but some contain infillings, such as calcite, which inhibit core breakage, but, during excavation, the mass will break almost as readily on these 'healed' fractures as on the clean fractures. Therefore, while RQD is a good measure of rock quality, it must be interpreted with great care by both geologist and engineer.

Although designed primarily for the interpretation of rock quality based on drilling cores, RQD has been used to assess the quality of rock in exposures and excavations. In this case, RQD reduces almost directly to a measure of fracture spacing only, since traverses to determine RQD measure the lengths between fractures which are 10 cm or more, and the sum of these is expressed as a percentage of the traverse length. This seems to have little to commend it in relation to a normal fracture analysis.

3.10 Mechanical classification of the rock mass

The classification of a rock mass for engineering purposes is difficult; this is due to the number of parameters involved, and to the diversity of interests of the engineer. It has been stated above that the term 'quality' can be interpreted in terms of the engineering use to which the rock mass is to be placed. The situation is similar for a classification: one designed primarily for dam engineering would have little value for tunnelling, and one designed for building would have little value for either dams or tunnels. No all-purpose mechanical or engineering classification of the rock mass has been developed. Here, it is possible to discuss general ideas only.

Initially, the distribution of rock types, the nature and regional pattern of structural discontinuities, weathering patterns, and the existence and origin of residual stress fields, will all be determined by a regional geological survey to provide fundamental data. These data provide a general pattern. Detailed examination of outcrops, excavations and bore cores, together with field testing of the rock mass and laboratory testing of the rock material will be required to give adequate data for any classification which might be devised. Finally, empirical data from any works already completed in the rock mass may be used for classification purposes. In fact, most of the classifications currently in use are based, to a large extent, on such empirical data, usually from successful undertakings. Rarely is the case of failure considered. Existing classifications have been developed for a specific purpose such as tunnelling, or rock slopes. Rarely, if ever, is it possible to extend a specific-purpose classification to a more general use.

One of the first attempts to develop a rock mass classification was made by Terzaghi for tunnelling purposes. This was a descriptive–qualitative classification, and while it was not only useful for the purpose for which it was designed and was capable of application to concrete dam foundations, it is now rarely used. The first system using quantitative data was the Salzburg classification which used as its basis the strength of the rock material; the orientation, spacing, condition, and coefficient of friction of joints; the presence of water; and the size of the tunnel. Deere has proposed RQD as a general purpose classification, but this lacks a great deal of important information.

Probably the most satisfactory classification is that due to Bieniawski (1975), which is known as the South African Geomechanics Classification (SAGC).

3.10.1 THE SOUTH AFRICAN GEOMECHANICS CLASSIFICATION

A rock mass is divided into a number of units, each having a characteristic set of properties, i.e. it is divided into homogeneous units, with the same rock type, and uniform fracture pattern and fracture spacing. For each unit, the following parameters are determined:

(i) uniaxial compressive strength;
(ii) RQD;
(iii) weathering characteristics;
(iv) joint and other fracture spacing;
(v) fracture openness, continuity and infilling;
(vi) orientation of fractures;
(vii) groundwater conditions.

On the basis of these criteria, five classes may be recognized. Relative indices of various parameters are given in Table 3.8. Since the parameters are considered to be of unequal importance, each is assigned a weighted minimal value, called a rating. The sum of all of the values of the various parameters defines the rock quality.

Although this classification was designed primarily for underground mining, it is applicable to rock slopes and foundations. It should be noted that while homogeneity and continuity are considered to be important, no account is taken of anisotropy or elasticity, both of which are important considerations, especially in determining the stress distribution around openings. The tensile strength of the rock material is another parameter not taken into account. This can be extremely significant in mining operations.

3.10.2 ROCK STRUCTURE RATING AND ROCK MASS QUALITY

These two classifications were also developed for use in mining. The former, due to Wickham and Tiedmann, is based on case history studies for a variety of rock types. It considers three basic parameters:

(i) rock type, folding and discontinuities;
(ii) joint spacing and orientation;
(iii) water inflow and joint openings.

Values are assigned to each parameter, and the 'rock structure rating,' RSR, is computed as follows:

$$RSR = (i) + (ii) + (iii)$$

The designers of this system have found a good correlation between RSR and the rock loads on steel supports in tunnels. The higher the value of RSR, the lower the rock loads, and hence a reduction in support requirements.

The Rock Mass Quality system was devised by Barton *et al* (1974), again for underground excavation. Six parameters are considered:

(i) RQD;
(ii) joint set number J_n;
(iii) joint roughness number J_r;

Table 3.8 Geomechanics classification of jointed rock masses. (After Bieniawski 1975.)

(a) *Classification parameters and their ratings*

	Parameter						
1	Strength of intact rock material	Point-load strength index (MPa)	>8	4–8	2–4	1–2	Use of uniaxial compressive test preferred
		Uniaxial compressive strength (MPa)	>200	100–200	50–100	25–50	10–25 / 3–10 / 1–3
	Rating		15	12	7	4	2 / 1 / 0
2	Drill core quality RQD (%)		90–100	75–90	50–75	25–50	<25
	Rating		20	17	13	8	3
3	Spacing of joints		>3 m	1–3 m	0.3–1 m	50–300 mm	<50 mm
	Rating		30	25	20	10	5
4	Condition of joints		Very rough surfaces Not continuous No separation Hard joint wall rock	Slightly rough surfaces Separation <1 mm Hard joint wall rock	Slightly rough surfaces Separation <1 mm Soft joint wall rock	Slickensided surfaces or Gouge <5 mm thick or Joints open 1–5 mm Continuous joints	Soft gouge >5 mm thick or Joints open >5 mm Continuous joints
	Rating		25	20	12	6	0
5	Ground water	Inflow per 10 m tunnel length	none	or 1 <25/min	or 1 25–125/min	or 1 >125/min	
		or Ratio (joint water pressure / major principal stress)	0	0.0–0.2		0.2–0.5	>0.5
		or General conditions	Completely dry	or Moist only (interstitial water)		or Water under moderate pressure	or Severe water problems
		Rating	10	7		4	0

continued on p.50

(b) *Adjustment for joint orientations*

Strike and dip orientations of joints		Very Favourable	Favourable	Fair	Unfavourable	Very unfavourable
Ratings	Tunnels	0	−2	−5	−10	−12
	Foundations	0	−2	−7	−15	−25
	Slopes	0	−5	−25	−50	−60

(c) *Rock mass classes and their ratings*

Class no.	I	II	III	IV	V
Description	Very good rock	Good rock	Fair rock	Poor rock	Very poor rock
Rating	100–90	90–70	70–50	50–25	<25

(d) *Meaning of rock mass classes*

Class no.	I	II	III	IV	V
Average stand-up time	10 years for 5 m span	6 months for 4 m span	1 week for 3 m span	5 h for 1.5 m span	10 min for 0.5 m span
Cohesion of the rock mass (kPa)	>300	200–300	150–200	100–150	<100
Friction angle of the rock mass	>45°	40–45°	35–40°	30–35°	<30°
Caveability of ore	Very poor	Will not cave readily Large fragments	Fair	Will cave readily Good fragmentation	Very good

(iv) joint alteration number J_a;

(v) joint water reduction factor J_w;

(vi) stress reduction fctor, SRF.

'Rock mass quality', RMQ, is given by: $RMQ = \frac{RQD}{J_n} \cdot \frac{J_r}{J_a} \cdot \frac{J_w}{SRF}$

This seems to be a complicated system, and of little value outside of the purpose for which it was conceived.

It can be claimed that the SAGC system is of more general applicability than any other system presently available. It is relatively simple, and partly quantitative. It must be noted that all systems are applicable primarily to hard, brittle, jointed rock masses. They would certainly give conservative results if applied to other rocks.

It is desirable that engineering geology should be trending towards quantification. However, as quantative and semiquantitative systems of classification of rock masses are examined, and applied, it is clear that they do not, in any sense, replace the sound, qualitative assessment of the experienced geologist.

References and further reading

Attewell P. B. & Farmer I. W. (1976) *The Principles of Engineering Geology*. Chapman & Hall, London.

Anon. (1977) The description of rock masses for engineering purposes. *Quart. Jl. Engng Geol* **10,** 355–88.

Barton N., Lien R. & Lunde J. (1974) Engineering classification of rock masses for the design of tunnel support. *Rock Mechanics* **6,** 189–236.

Beavis F. C. (1949) The geology of Clover Dam. *Min. Geol. Jl., Vict.*

Beavis F. C. (1962) The geology of the Kiewa Project. *Proc. Roy. Soc. Vict.* **75,** 349–410.

Bieniawski Z. T. (1973) Engineering classification of jointed rock masses. *Trans. Sth. African Inst. Civ. Eng.* **15,** 335–44.

Deere D. U. (1978) Geological considerations. In *Rock Mechanics in Engineering Practice* (eds K. G. Stagg and O. C. Zienkiewicz). John Wiley, London.

Fecker E. (1978) Geotechnical description and classification of joint surfaces. *Bull. I.A.E.G.* **18,** 111–20.

Fecker E. & Rengers N. (1971) Measurement of large scale roughness of rock planes. In *Proc. Int. Symp. Rock Mechs,* Nancy, pp. 1–18.

Hills E. S. (1945) *Outlines of Structural Geology*. Methuen, London.

Hobbs B., Means W. D. & Williams P. F. (1976) *An Outline of Structural Geology*. John Wiley, New York.

Jaeger J. C. (1956) *Elasticity, Fracture and Flow*. Methuen, London.

Price N. J. (1966) *Fault and Joint Development in Brittle and Semi-brittle Rock*. Pergamon Press, London.

Ragan D. M. (1973) *Structural Geology: An Introduction to Geometrical Techniques*. John Wiley, New York.

Terzaghi K. (1948) *Rock Tunnelling with Steel Supports*. Commercial Stamping and Shearing Co., Ohio.

4

Rock Weathering

4.1 Introduction

Rock weathering results from a series of processes which produce changes in the physical and mineralogical nature of both the rock material and the rock mass. These changes are gradual, and usually gradational; they are normally accompanied by a deterioration in mechanical properties, which becomes more pronounced the more advanced the changes due to weathering. Most rocks exposed at the surface are the products of processes which involved elevated pressure and/or elevated temperature. Since physical and chemical conditions at the surface are significantly different from those under which the rock was formed, the fabric and minerals of the rock undergo changes in response to the new physico-chemical regime.

Probably the most generally accepted definition of weathering is 'those processes of alteration of rocks under the direct influence of air and water'. Weathering is usually a slow process, although very rapid weathering has been recorded. The processes are, for the most part, complete only at and very close to the surface, although complete weathering to depths of 300 m has been recorded from north-east Victoria, Australia, and at up to 1500 m in the USSR. Just as depth of weathering may vary from place to place, patterns of weathering can be extremely variable even in an apparently homogeneous rock mass. The nature of weathering itself, and the control of depth to which rocks weather and of weathering patterns, are problems about which a great deal remains to be learned.

4.2 Weathering processes

Rocks weather in two principal ways: by physical *disintegration* and by chemical *decomposition*. Physical weathering involves the mechanical breakdown of the rock mass (block disintegration, Fig. 4.1), usually controlled by discontinuities in the mass (Fig 4.2); and mechanical breakdown of the rock material (granular disintegration) controlled by micro-discontinuities such as grain boundaries and mineral cleavages. Chemical decomposition affects almost all minerals; only a few, notably quartz, are more or less immune. The processes involved in decomposition are oxidation, with or without changes in valency; reduction; hydration; hydrolysis; carbonation; and solution.

Under temperate humid climatic conditions, decomposition and disinte-

Fig 4.1 Block disintegration due to weathering on joints and bedding planes in siltstone, Coff's Harbour, New South Wales, Australia.

Fig 4.2 Sheeting exfoliation in granite, Moonbi, New South Wales, Australia.

gration proceed simultaneously and it may be difficult to isolate the direct effects of each. The relative importance of the two groups of processes is very much a function of climate. In arid and frigid regions, disintegration tends to be dominant; in humid tropical zones, decomposition is far more effective than disintegration, although this latter occurs as a result of decomposition. The importance of disintegration is that it opens up both the mass and material so that access is available for agents of decomposition.

Most authorities recognize a third group of processes: the biological, due to the physical and chemical effects of animals and plants (Table 4.1). Biological attack may or may not be significant, but it is true that a close relationship exists between type and density of vegetation, and the degree of weathering. Plants, and animals, provide organic material which will react with rocks to produce decomposition.

The rate at which the various processes operate, i.e. the rate at which weathering takes place, is a function of three sets of conditions:

(i) environmental; dominated by climate and material, but also involved are topography, hydrological conditions and the biological systems;

(ii) the properties of the rock mass, especially the homogeneity, and the nature, spacing and pattern of discontinuities, i.e. the macrofabric;

(iii) the properties of the rock material, including composition, fabric, texture and permeability.

Table 4.1 Processes of rock weathering.

Type of weathering	Processes	Effects
Physical disintegration	Sheeting Volume change Frost plucking Freezing and thawing Heating Colloid plucking Abrasion	Particle size reduction resulting in increased surface area. No chemical changes (1) Block disintegration (2) Granular disintegration,
Chemical decomposition	Solution, hydration, hydrolysis Oxidation Reduction Carbonation	Production of new minerals and loss of mineral material Changes in fabric
Biological weathering	Physical abrasion by burrowing animals Wedging by plant roots Extraction of nutrients	Physical, chemical and biochemical effects Bacterial activity Disintegration Decomposition

4.2.1 CLIMATE AND WEATHERING PROCESSES

The relationship between climate and weathering processes is shown in Fig. 4.3.

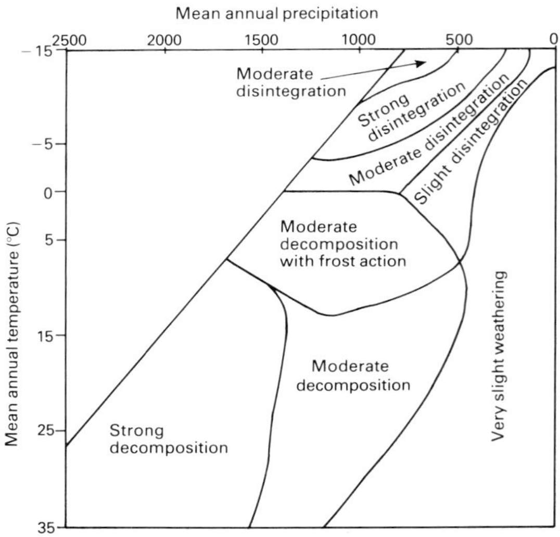

Fig 4.3 Postulated relationship between climate and type of weathering. [After Fookes *et al* (1971).]

This illustrates the role of climate in determining which process, or group of processes, are effective in the weathering processes. In practical terms, however, care must be exercised in relating observed weathering and present climate. It is very probable that, in view of the slow rate of weathering, and the relative rapidity of climatic change in the Pleistocene–Quaternary period, some of the observed weathering is due to past climates. For example, weathering in at least part of the arid zone of Australia is the result of early Quaternary humid conditions, and not of the present arid climate.

Climate influences weathering both directly and indirectly. Temperature is important in controlling the rate of chemical reactions, while it also acts physically in terms of frost action, and heating and cooling, leading to physical disintegration, due to fatigue. Because higher temperatures accelerate reactions, chemical decomposition is important in hot, humid tropical regions, and of special importance is the solution of silica under these conditions. In arid and frigid regions chemical decomposition is inhibited: in the former due to lack of water; in the latter depressed temperatures are largely responsible. In both of the arid and frigid climatic zones, abrasion is an important physical process. Wind-blown sand in arid regions and ice-borne rock debris in glacial areas are particularly important abrasive agents.

Precipitation and the precipitation–evaporation ratio influence the products of weathering. If precipitation exceeds evaporation, ions are lost in both surface and subsurface run-off (solution). If evaporation exceeds precipitation, then salts accumulate, and such minerals as halite, gypsum and calcite are to be found in the weathered rock (Fig. 4.4). Clearly if the precipitation–evaporation ratio

Fig 4.4 Scanning electron micrograph of fine stellate crystals of gypsum deposited on quartz and illite as a result of weathering of a shale under arid conditions (× 5000) Fowlers Gap, New South Wales, Australia.

becomes extremely low, as in all arid regions (0.005 or less) chemical weathering may be absent, except for salt weathering, which is enhanced.

4.2.2 TOPOGRAPHY AND WEATHERING

Although it is certain that a relationship exists between landform and both the rate and pattern of weathering, the exact nature of this relationship is not always clear. It is known that, for granitic masses in south-eastern Australia, as a general rule, in deeply dissected terrains, weathering tends to be deepest and most advanced on plateaux and ridges, and to be slight, and shallow, low down on valley walls, while fresh rock may outcrop in stream beds (Figs 4.5a, b). Similar patterns have been reported from Hong Kong (Figs 4.5c,d).

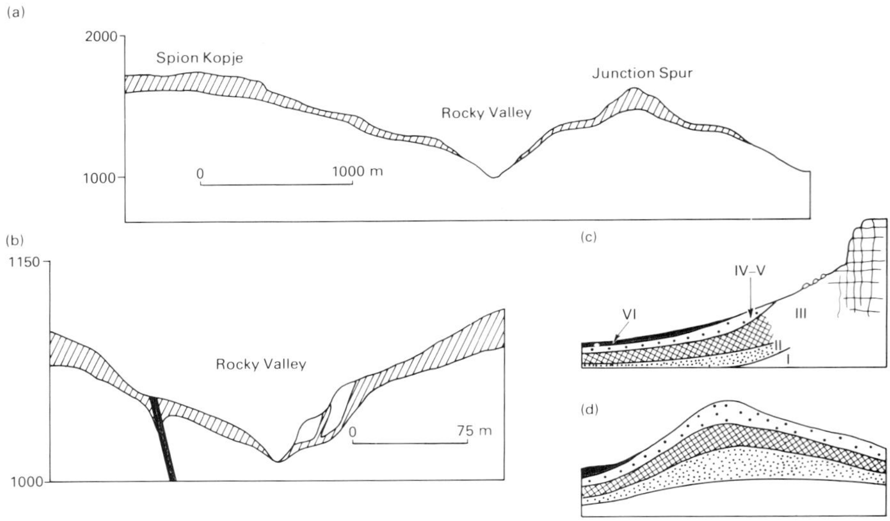

Fig 4.5 Weathering patterns: (a) General pattern of weathering of a granitic mass, Kiewa Australia. (b) Detail of weathering pattern in Rocky Valley. (c) Weathering pattern on a retreating cliff. [After Ruxton and Berry (1957).] (d) Weathering pattern on a low hill. [After Ruxton and Berry (1957).]

The influence of landform on weathering is due to a combination of hydrological and geomorphic factors. Topography, to a considerable degree, controls the rate of run-off and hence the rate of water intake to the rock mass. Topography also influences the rate of subsurface drainage and so the rate of leaching out of soluble constituents. The rate of erosion of completely weathered rock, and the consequent rate of exposure of less weathered rock, to agents of weathering is also a function of topography. On steep slopes, the run-off coefficient is high: very little water penetrates into the rock mass. At the same time, on such slopes, erosion is extremely active. As a consequence, disintegration tends to be more effective than decomposition. Flat areas have little or no run-off: infiltration is high, and erosion is low to zero. Chemical decomposition is here extremely active while disintegration is a minor process.

In considering the relationship between topography and weathering, the geomorphic history must be understood, because the observed weathering may be associated with more than one phase of topographic development. In the example shown in Fig. 4.5a, the very deep weathering is associated with an old (probably Tertiary or even Mesozoic) flat land surface, while the shallow weathering is associated with a currently developing land surface. In assessing weathering patterns both past climates and past landforms have to be taken into account.

4.2.3 WEATHERING AND FABRIC

Weathering rate and degree of weathering are functions of the rock fabric: both the fabric of the rock mass (the macrofabric), and the fabric of the rock material (the microfabric). The more closely the fractures, or other discontinuities, are spaced, the greater will be the surface area exposed to weathering agents, and so these agents will be increasingly effective. This point, at the scale of the rock mass, is illustrated by an excellent example from Dartmoor, England (Fig. 4.6).

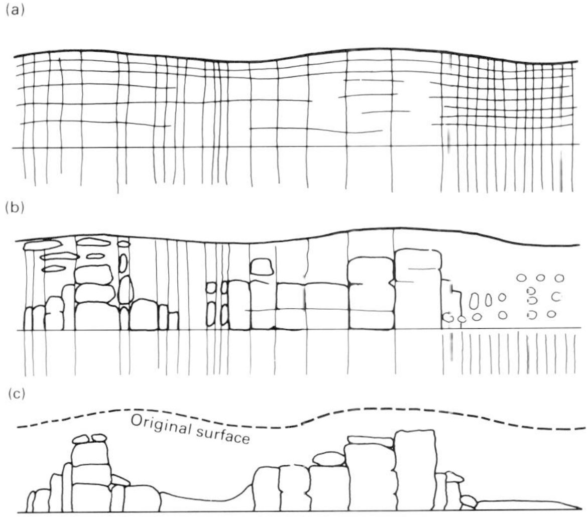

Fig 4.6 Hypothetical development of terrain on a granite mass due to weathering and erosion [After Fookes *et al* (1971).]

It is well known, also, that weathering tends to progress more rapidly along faults and fault zones where, due to cataclasis, the rock has been crushed, so that in a sense, physical disintegration is complete even before any chemical weathering of the rock mass begins. Two Australian examples are shown in Fig. 4.7; the Wyangala Dam, New South Wales, and the McKay Underground Power Station, where abnormally deep weathering is associated with faults, and with zones of closely spaced joints.

Good correlation frequently exists between fracture spacing and degree of weathering. However, it is very easy in this case to come to false conclusions. While fracturing facilitates weathering, weathering itself induces fracturing. The question has always to be asked: which is cause, and which effect?

As macrofabric of the rock mass influences both the nature and rate of weathering, so also does the microfabric of the rock material influence the nature and rate of weathering of the material. Weathering also leads to changes in the microfabric, even if the macrofabric (Fig. 4.8) is preserved. The changes induced in the microfabric of a granite with weathering are largely responsible for the changes in the engineering properties of the rock material. Grain boundaries, mineral cleavages, and intergranular pore spaces allow weathering agents to penetrate the rock material. As weathering progresses, not only are these fabric

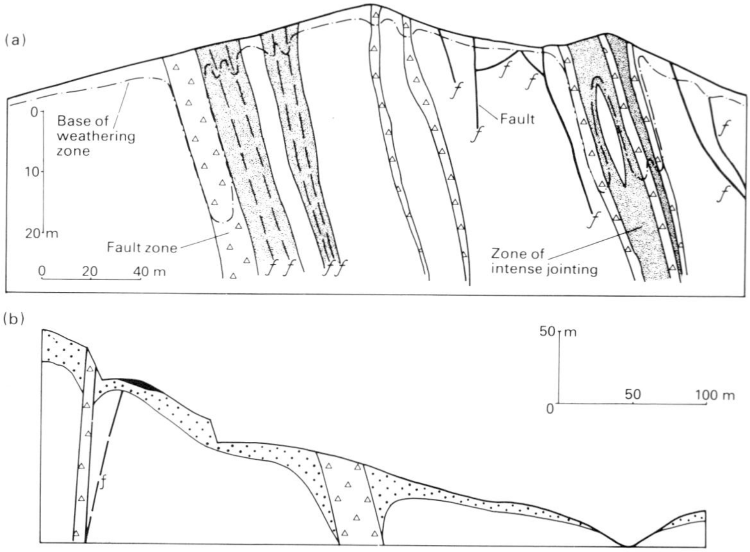

Fig 4.7 (a) Weathering on faults, Wyangala Dam, Australia [After Thomson (1971).] (b) Weathering in a faulted granite mass, Kiewa 1 Underground Power Station, Australia.

Fig 4.8 Preservation of macrofrabric in completely weathered biotite sillimanite gneiss Cranky Charlie, Victoria, Australia.

elements emphasized, but new microfractures and voids are developed. The enlarging of existing microdefects and the initiation of new defects are probably due primarily to solution (Fig. 4.9)

Rocks with a fine, compact fabric, in which adjacent crystals are welded or cemented together, have the greatest resistance to the entry of water and of solutions. Those with coarse or open fabrics, especially if these are combined with high porosity, offer the least resistance (Fig. 4.10). Coarse sandstones and granitoid rocks therefore tend to be more accessible to solutions than fine shales and volcanic rocks. A direct relationship exists between weathering susceptibility and permeability: both are highly dependent on microfabric.

Changes in microfabric with weathering are now being studied using scanning electron microscopy, and some useful and interesting results are being

Fig 4.9 Scanning electron micrograph of solution of calcite crystal in weathered shale (× 1500), Fowlers Gap, New South Wales, Australia.

obtained. These studies have shown that for granitic rocks the initial ingress of weathering agents occurs along primary microvoids and mineral cleavages, and that the initial stages of weathering increase porosity by solution along grain boundaries and, in granites, within feldspars. Weathering greatly increases the

Fig 4.10 Cavernous weathering of a sandstone, Wollombi, New South Wales, Australia.

intensity of microfracturing by opening grain boundaries, expanding biotite crystals, and possibly by destressing quartz crystals. Continued weathering of feldspars produces a variety of microfabric elements, including an extremely open porous feldspar microfabric; and a clay microfabric varying from dense, tightly packed aggregates, to extremely porous open microfabrics, and delicate porous networks of particles.

4.2.4 WEATHERING AND MINERAL COMPOSITION

It is known that some minerals are more stable with respect to weathering than others. Table 4.2 shows the relative stability of the common silicate minerals, ranging from the least stable olivine and Ca plagioclase, to the most stable rock forming mineral, quartz, which is resistant to chemical weathering under all but a few very special conditions.

Table 4.2 Stability of rock-forming silicate minerals with respect to decomposition.

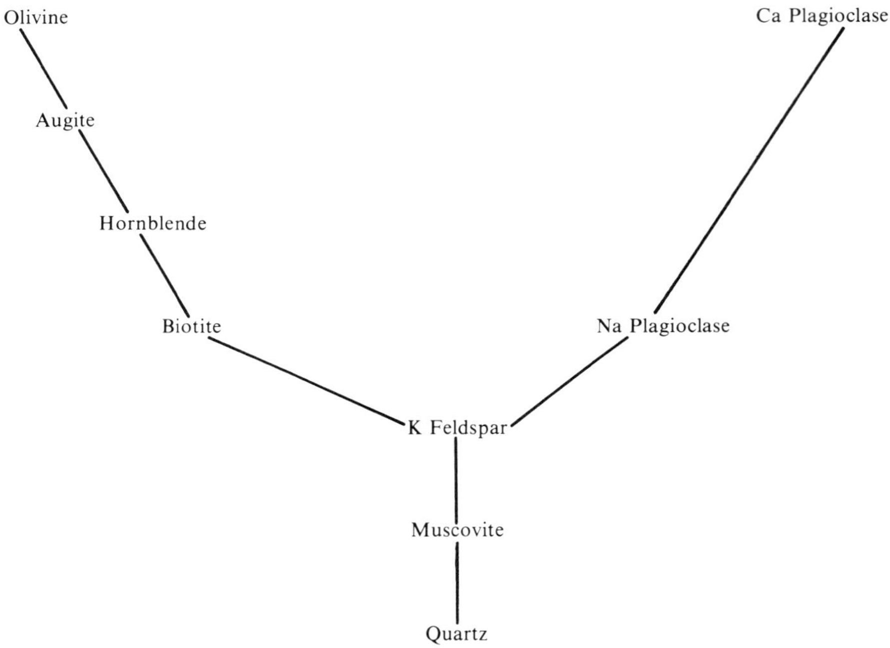

The sequence shown in Table 4.2 has led to both oversimplification and erroneous interpretation, especially by engineers. This is because when it is applied to the rock material, the overriding importance of fabric is neglected. Given a constant climatic factor, and constant fabric, then a gabbro, for example, composed essentially of olivine, augite and Ca plagioclase, will weather much more rapidly than a granite, composed essentially of quartz, K feldspar, and biotite. Geologists have often observed that a basalt is much more stable with respect to weathering than a granite, although it is composed essentially of the same minerals as gabbro. This is due to fabric differences. The coarse, open fabric of granite permits more ready access to agents of decomposition than does the fine, close fabric of basalt. Both laboratory studies and field obser vation suggest most strongly that, in terms of stability with respect to chemical decomposition, fabric is at least as significant as composition, if not more so.

A very simple explanation for the stability variations of the rock-forming silicate minerals is sometimes presented, the basis of which is the close parallel between the stability sequence of Table 4.2 and the so-called reaction series of petrology. It is argued that olivine and Ca plagioclase, being the first minerals to crystallize out from a magma are, in the exposed rock, further from their stability pressure–temperature range than is quartz, which is the last mineral to crystallize. Such a simple explanation, while possibly containing some elements of truth, obscures the basic controlling factors.

The importance of mineral structure in determining the relative rates of chemical weathering of silicate minerals cannot be overemphasized. For example, olivine consists of isolated silica tetrahedra bonded together by Mg^{2+} and Fe^{2+}. Both of these cations are readily leached from the margins and fracture surfaces of crystals; the loss of the cationic 'bridges' releases the tetrahedra, and, as a consequence, the weathering of olivine proceeds rapidly. On the other hand, the Na and K feldspars are relatively resistant to decomposition because the framework structure of the tetrahedra militates against the escape of K^+ and Na^+, despite the fact that these ions are highly mobile. To free these cations, the tetrahedral chains must be broken. Apparently the Al–O–Si bonds form the weakest links in the chains, and the greater the degree of substitution of Al for Si in the tetrahedra, the greater the number of weak links. Hence, the considerable range in weathering stability of the feldspar group.

4.3 Patterns of weathering

The pattern of rock weathering at an engineering site is often one of the most important of the geological factors to be considered by engineers and engineering geologists. In Sections 4.2.2 and 4.2.3 some general patterns were noted, but, at the scale of a site investigation, or at even larger scales, major departures from these idealized general patterns almost invariably occur.

One of the most firmly held beliefs is that weathering is complete at the surface, and that it decreases gradually and progressively with increasing depth. If the classification to be discussed in Section 4.5 is used, then the weathering sequence would be that shown in Table 4.3.

Table 4.3 Ideal sequence of weathering grades.

Surface	Soil
↓ Increasing depth	Completely weathered rock (CW)
	Highly weathered rock (HW)
	Moderately weathered rock (MW)
	Slightly weathered rock (SW)
	Fresh rock (F)

Table 4.4 shows some actual conditions observed from cores of bores drilled through the weathered zone into fresh rock. Both bores were sunk in the same

Table 4.4 Patterns of weathering observed in bore cores.

Bore A		Bore B	
Depth (m)	Weathering grade	Depth (m)	Weathering grade
0–11	CW	0–5	HW
11–14	HW	5–14	MW–SW
14–19	MW	14–19	MW
19–23	SW	19–25	HW–MW
23–27	CW	25–27.5	MW–SW
27–30	SW	27.5–31	MW
30–34	HW–SW	31–38	SW
34–65	F	38–75	F

general area, about 300 m apart, one in dolomitic phyllite (A) and one in calcareous phyllite (B). If the rock were homogeneous with respect to composition and with respect to distribution and spacing of discontinuities, the ideal sequence would be a possibility. Where variations in rock type, and in the nature, spacing and orientation of fractures occur, departures from the ideal weathering pattern must be expected. It is these two factors: compositional homogeneity of the rock mass, and the nature, spacing, orientation and condition of discontinuities which will determine the pattern of weathering of the rock mass. Figure 4.11 shows some weathering patterns, based on detailed

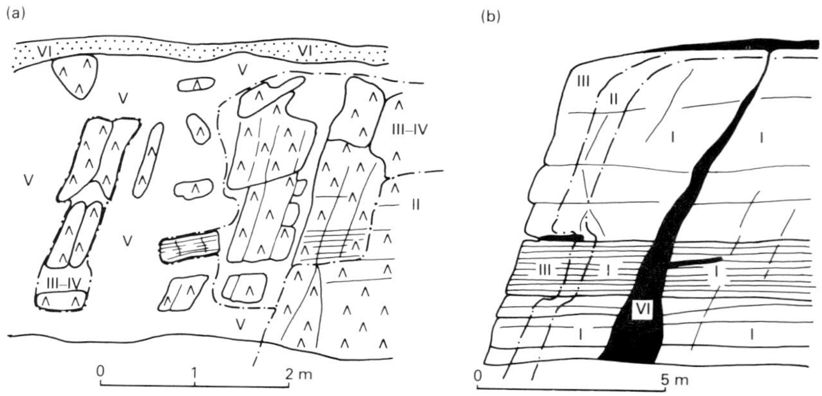

Fig 4.11 Pattern of weathering in exposed rock faces: (a) Basalt, Illawarra Highway, Wollongong, New South Wales, Australia. (b) Sandstone and shale, F5 Tollway, Waterfall, New South Wales, Australia.

studies, for several types of rock mass. The importance of rock type, and fracturing, is clearly seen.

Boundaries between different grades of weathering may be quite sharp, or they may be gradational over a distance of several metres. In a homogeneous rock mass, the weathering does tend to be gradational: however, if different lithologies are in contact, the change from one grade of weathering to another may be sharp and clearly defined. Often, it is found both in bore core and in exposures that various grades of weathering are apparently randomly intermixed (Figs 4.12, 4.13). All of these frequently observed complexities show the critical

Fig 4.12 Slightly weathered core stones in a mass of completely weathered tonalite, West Kiewa, Victoria, Australia.

significance of sometimes seemingly minor variations in composition, texture, and fabric of the parent rock.

It is sometimes observed that soils in the weathering profile which were obviously formed *in situ* are developed directly on moderately weathered rock, with highly and completely weathered sections of the profile absent. Climatic influences may be responsible for this anomaly, which is most commonly observed in tropical lateritic soils, although it is sometimes observed in temperate zone, and hot arid zone, profiles.

4.4 Products of weathering

All rocks weather ultimately to a soil, and the types of the soils so produced are of considerable importance to the engineer. The nature and composition of weathered rocks and soils are a function of the composition of the parent rock and of climate. If a rock, under extreme arid, or extreme frigid, conditions is subjected to purely physical disintegration, the final product of weathering will be a soil with a mineral composition almost exactly that of the parent rock. If, however, chemical decomposition has been involved, the mineral composition of the soil will differ strongly, or even completely, from that of the parent rock.

All rock-forming silicate minerals, with the exception of quartz, will weather, ultimately, to clay minerals, under the influence of chemical agents (Table 4.5). Carbonate minerals undergo partial to complete solution. It can be seen from Table 4.5 that the clay mineral formed does not depend solely on the parent mineral: other factors are involved, especially climate, but also the availability of ions of certain types.

The importance of climate can be appreciated if two extremes are studied: weathering in a hot arid climate, and weathering in a humid tropical climate.

In hot arid regions, not only does evaporation exceed precipitation by a large factor, but also, since precipitation events are brief, and widely spaced in time, any water which enters the rock mass is lost very quickly by evaporation. As a

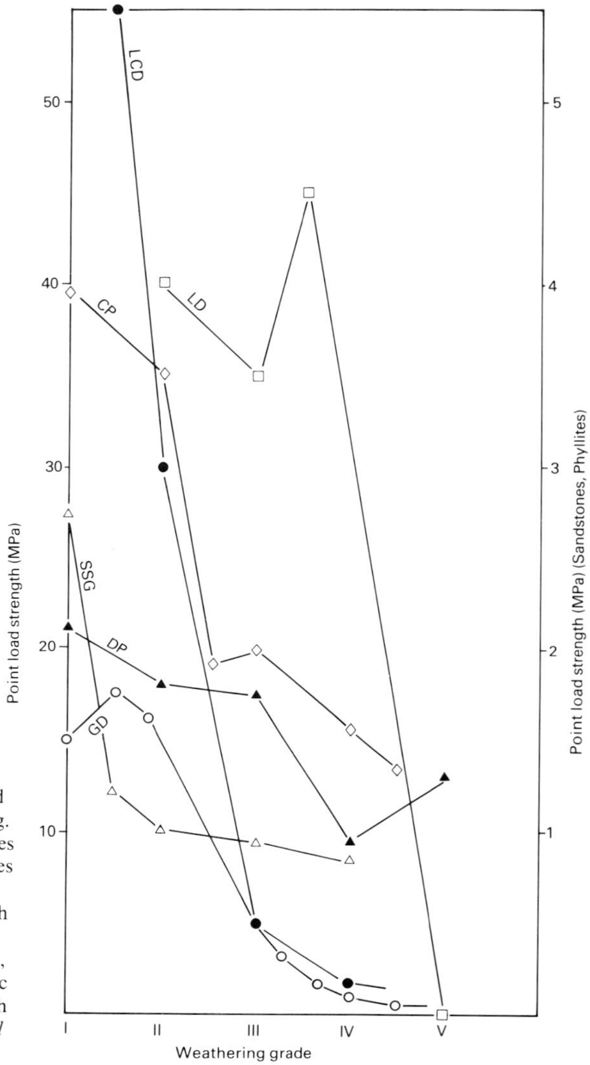

Fig 4.13 Variation of point load strength with grade of weathering. LCD, limestone and chert (Fookes *et al* 1971). LD, limestone (Fookes *et al* 1971). CP, Calcareous phyllite, Fowlers Gap, New South Wales, Australia. SSG, Hawkesbury Sandstone, Gosford, New South Wales. DP, Dolomitic phyllite, Fowlers Gap, New South Wales. GD, Granite (Fookes *et al* 1971).

result, soluble constituents are not removed, and reactions are retarded. Oxidizing conditions prevail, and iron is retained as Fe^{3+}. The characteristic clay minerals produced are montmorillonite, illite and chlorite, or mixed layers of these minerals. Unless present in the parent rock, kaolinite is absent. If it is present in the parent rock, it remains unaltered.

By contrast, rocks in humid tropical areas are strongly leached due to the continual movement of downward percolating water, and the soluble products of weathering reactions are removed immediately they are formed. Chemical weathering proceeds rapidly by the loss of these constituents, while the residue becomes progressively enriched in minerals which contain a high proportion of

Table 4.5 Weathering products of silicate minerals (after Loughnan).

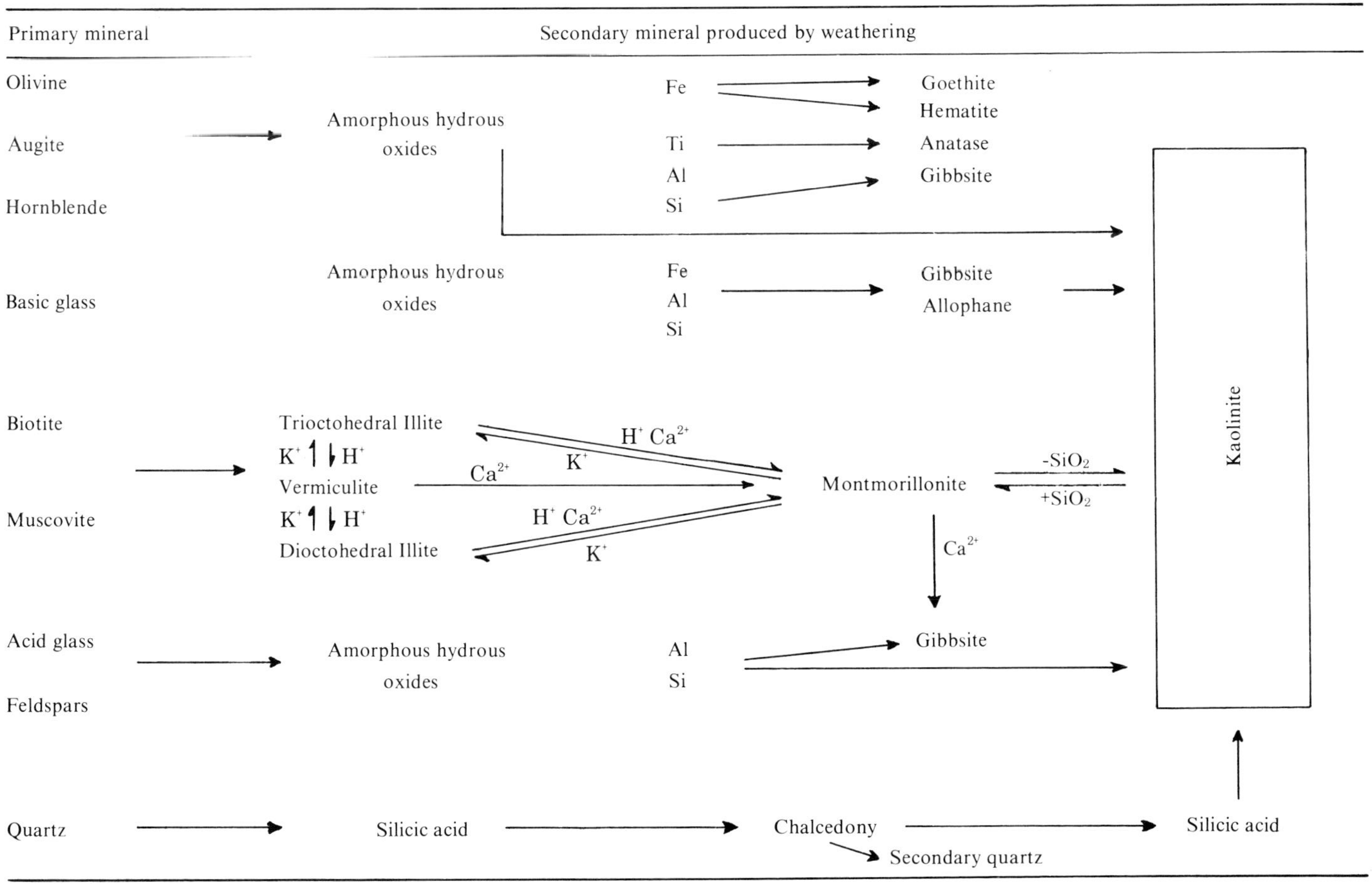

Al_2O_3, such as kaolinite, halloysite, gibbsite and boehmite. Three examples of the mineralogy of completely weathered pelitic rocks of similar composition, but weathered under three different climatic conditions, are presented in Table 4.6.

Table 4.6 Mineral composition of completely weathered pelitic rocks.

Location	Broken Hill	Weipa	Goulburn
Climate	Hot arid	Humid tropical	Humid temperate
Mineral			
Quartz	28	8	50
Illite	32	—	24
Chlorite	15	—	—
Montmorillonite	12	—	7
Kaolinite	—	12	14
Gibbsite	—	68	—
Goethite	5	10	—
Gypsum	5	—	—
Calcite	2	—	—
Others	1	2	5

Since climate plays such an important role in rock weathering, it is impossible to make broad generalizations about the composition and fabric of any particular weathering product. However, provided the climatic factor is constant, it is possible to assess the consequence of the type of parent rock on the weathered product. For example, at Bathurst, Australia, a basalt, composed of Ca–Na plagioclase, olivine, titanaugite, magnetite and basic glass is, in the completely weathered condition, an expansive clay soil, composed of montmorillonite, with lesser kaolinite and halloysite. Some goethite and hematite are also present. Granite in the same area has completely weathered to a well-graded sandy silty clay composed of quartz, fine mica, fragments of feldspar, kaolinite, halloysite and chlorite.

4.5 Classification of weathered rock for engineering purposes

While rock names, if used alone, can be misleading as indicators of geomechanical and engineering properties, the situation is even more hazardous if weathered rocks are involved. A classification of weathered rocks for engineering purposes requires as its basis, characteristics of the rock which can be determined in the field, or quite readily in the laboratory. Classifications range from those based purely on visual inspection, and which are purely subjective, to those based on the results of precise and detailed testing. The most generally used classifications are based on characteristics which can be determined by visual inspection together with the results of a number of relatively simple index tests.

The classification most frequently used, due to Fookes *et al* (1971) as modified by Dearman (1974, 1976), was designed for granitic rocks from Dartmoor, England. While the authors quite specifically cautioned against its use for other than granitic rocks, both geologists and engineers have applied it to all sorts of rocks, in all parts of the world, sometimes with most undesirable results. The classification shown in Table 4.7 is satisfactory for granitic and crystalline metamorphic rocks. It is not always suitable for many volcanic rocks, sedimentary rocks, or low grade metamorphic rocks. Despite these comments, the principle of the classification is sound, and it is possible to apply the principles to construct a classification for any rock mass under consideration. Table 4.8 is a classification for clastic sedimentary rocks, illustrating the applications of the principles used by Dearman for granites to rocks of an entirely different composition and fabric.

One most undesirable practice, not uncommon amongst both geologists and engineers, is to classify the rock, in terms of weathering grade, purely on some simple, visual characteristic, such as colour. Index tests are then carried out and the results assigned to various colour grades, which are not necessarily weathering grades. It is essential that all tests be carried out before the rock is classified. Another undesirable practice is the classification of the rock mass weathering simply in terms of the rock material. Characteristics of both mass and material must be considered.

Under field conditions, few problems arise in the recognition of fresh rock, and of completely weathered rock. Separation of grades II, III and IV may present difficulties, and this is especially true for grades III and IV. Determination of the rock:soil ratio (RSR), i.e. the relative proportions of rock and soil, is an excellent field criterion for separation, but the final decision must be made on the basis of the results of index tests. A number of indexes may be used, but it is not always necessary to use all of the indexes. In some cases, it may be difficult to obtain values of a particular index: e.g. it is often very difficult, in sedimentary rocks, to determine the rock:soil ratio. In classifying weathered rock for engineering purposes, several (at least) of the following criteria should be used:

(a) description of rock, including fabric and colour (material and mass);
(b) strength index (Point load index I_{s50}) (material);
(c) fracture spacing index (If) (mass);
(d) rock quality designation (RQD) (mass);
(e) rock:soil ratio (mass);
(f) elasticity (material);
(g) porosity (material);
(h) micro-indexes (material);
  (i) micropetrographic index (I_{mp})
  (ii) microfracture index (I_{fr}).

4.5.1 STRENGTH INDEX OF WEATHERING

Weathering of rocks is almost invariably accompanied by a reduction in strength. Exceptional cases are those in which secondary cementation by silica, iron oxides, or carbonates occurs as a result of the weathering processes. For a particular rock, it is to be expected that, in general, an increase in the degree of weathering will result in a progressive reduction in strength (Fig. 4.13). The point load index test has proved to be an extremely useful and rapid means for

Table 4.7 Classification of weathered crystalline rocks (After Irfan and Dearman).

Class	Grade	Description of material and mass	Index properties			
			Strength	RQD (%)	Rock : soil (% rock)	Effective porosity
Fresh	I	No visible sign of material weathering. Near boundary with Grade II some slight discolouration on major defects	Very high	Very high 90–100	95–100	
Slightly weathered	II	Discolouration indicates weathering of rock material and defect surfaces.	Range from very high for IIi to 50–60% of fresh rock strength for IIiv.	75–90	90–95	5% increase from rock rock
a. Partly stained	IIi	Defect surfaces, only, discoloured				
	IIii	Discolouration penetrates inwards from defects. Less than 50% discolouration by volume.				
b. Completely stained	IIiii	More than 50% discolouration				
	IIiv	100% discolouration				
Moderately weathered	III	Less than 50% of material decomposed and disintegrated to a weakly cohesive friable intact soil. Core rock discoloured, weakened.	30% of fresh rock strength	40–75	60–90	7% increase from fresh rock
Highly weathered	IV	More than 50% of material decomposed and disintegrated to a weakly cohesive friable intact soil. Rock cores discoloured and weakened.	15% of fresh rock strength	10–40	30–60	10% increase from fresh rock
Completely weathered	V	Intact friable soil which may be weakly cohesive. Soil has fabric of parent rock.	Extremely low	0–10	0–30	20% increase from fresh rock
Residual soil	VI	Friable soil with original rock fabric completely destroyed.	Extremely low	0	0	>20

Table 4.8 Classification of weathered pelitic rocks in arid regions for engineering purposes. [Values tabulated are multiples of the Maximum Value (M).]

Class	Grade	Mass characteristics	Material characteristics					Remarks
		General	I_f	I_s	*UCS*	E_t	Porosity	
Fresh	I and I(i)	Joints and other fractures such as cleavage and bedding tight and unstained.	Maximum Value (M)	Maximum Value (M)	Maximum Value (M)	Maximum Value (M)	Minimum Value (M)	Some chemical weathering as indicated by montmorillonite and a vermiculite-like mineral. Clay minerals also include chlorite and muscovite. (Grade I(i). Rock grey in colour.
Slightly weathered	II	Joints and other major fractures stained. Cleavages show little or no staining. Fractures generally relatively tight.	0.6–0.5	0.75–0.6	0.9–0.4	0.6–0.5	1.15–1.5	Block disintegration begins in separation along joints. Clay minerals: chlorite, montmorillonite and muscovite. Colour of rock, grey.
Moderately weathered	III	Joints and bedding planes open and strongly stained, with weathering penetrating material adjacent to fractures.	0.5–0.3	0.6–0.4	0.4–0.2	0.5–0.3	1.5–1.75	Some solution of minerals, especially calcite, sulphides and iron oxides. Goethite finely dispersed. Clay minerals: chlorite, montmorillonite and muscovite. Colour of rock, grey-pale fawn.
Highly weathered	IV	Joints and bedding planes open with up to 3 cm of weathered material. Some weakening and distintegration on cleavage.	0.3–0.1	0.4–0.2	0.2–0.15	0.3–0.1	1.75–2	Characteristic 'surface' minerals appear: halite, gypsum, 'lime' (calcite). Strong solution of calcite, sulphides and iron oxides. Clay minerals: kaolinite, montmorillonite, chlorite, muscovite and illite. Rock light fawn to brown in colour.
Completely weathered	V	Strong disintegration on fractures, and granular disintegration advanced. Soil development with up to 40% of mass as soil. Rock breaks down readily under rollers.	<0.1	<0.2	<0.15	<0.1	>2	Sulphides and ferric oxides completely replaced by calcite or goethite. Clay minerals: kaolinite, montmorrillonite, muscovite and illite. Colour of rock, brown to red brown.
Soil	VI	Fragments of CW and HW rock in mass. Rock fragments 10e-20% of total mass.						Original fabric of rock completely destroyed. Halite, gypsum, 'lime' abundant. Clay minerals: kaolinite, 'degraded' illite, muscovite, illite, montmorillonite, and chlorite.

determining the relative strengths of weathered rocks in the field, provided a large number of tests are made. On the basis of such tests, Fookes *et al* (1971) devised a strength classification scale (Table 4.9).

Table 4.9 Strength classification scale.

Description	Point load index I_{S50}(MPa)
Extremely high strength	>10
Very high strength	3–10
High strength	1–
Medium strength	0.3–1
Low strength	0.1–0.3
Very low strength	0.03–0.1
Extremely low strength	<0.03

The point load test is an *index* of strength, which can be used, during field mapping or core logging, to indicate relative strengths. The test, however, is of

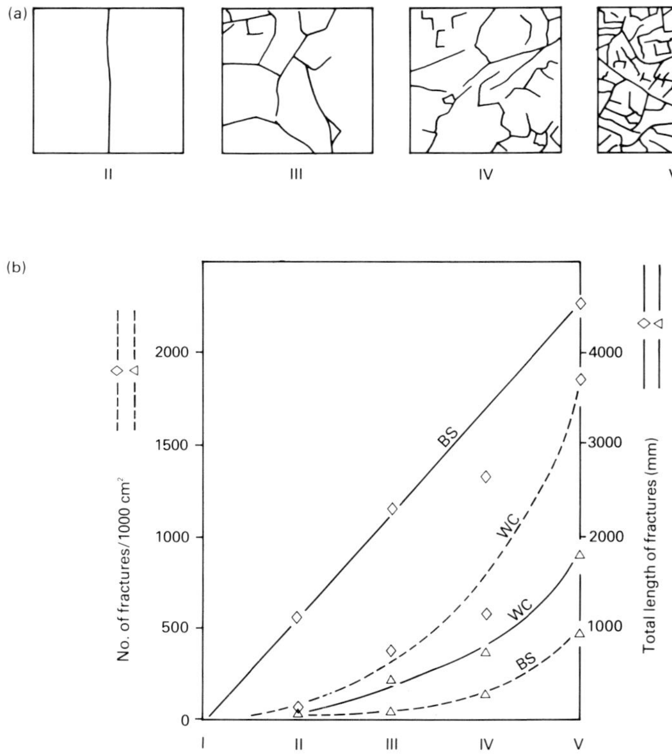

Fig 4.14 Fracturing and weathering in Bulgo Sandstone (BS) and Wombarra Claystone (WC), Wollongong, New South Wales. [After Evans (1978).] (a) Fracture patterns at grades II – V for Bulgo Sandstone. (b) Variation in fracture intensity with weathering.

little value for $I_{s50} < 0.1$ MPa; these values are usually recorded, in any case, in obviously completely decomposed rocks.

4.5.2 FRACTURE SPACING INDEX

This index is defined as the average spacing of fractures in cored or outcrop material. The spacing of fractures normally decreases with increased weathering; the fracture spacing index can therefore be a guide to the grade of weathering for a particular rock mass. Figure 4.14 shows the relationship between fracturing and weathering for two rocks from the Sydney Basin, Australia. In this case, the fracture spacing index as such was not used, but a measure was made of the number of fractures per unit area; the total length of fractures per unit area, which increased with increased weathering; and the mean length of fracture, which decreased with increased weathering. Figure 4.15 shows the fracture spacing index for various grades of weathering for the two phyllites from near Broken Hill, where it has been shown that fracture spacing influenced weathering, rather than the converse, since all of the fractures were demonstrably tectonic joints with the exception of closely spaced shrinkage cracks (2–3 mm) in the completely weathered dolomitic phyllite.

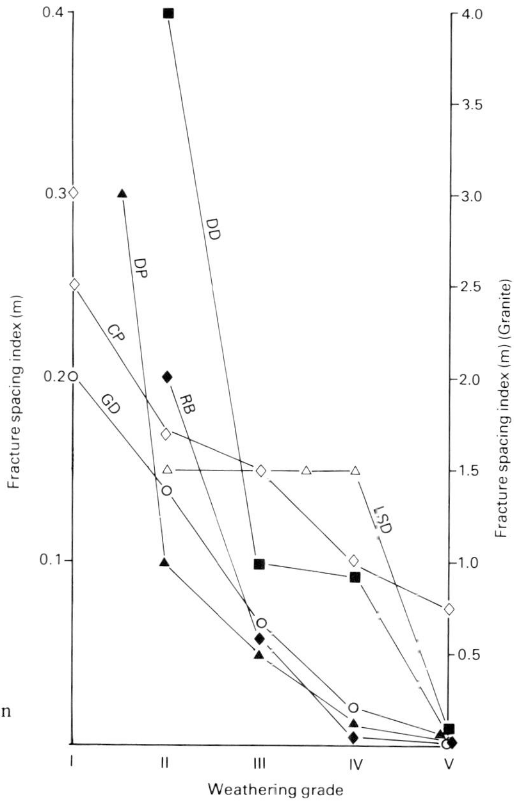

Fig 4.15 Variation in fracture spacing index with weathering. DD, Dolerite (Fookes *et al* 1971). LSD, Limestone (Fookes *et al* 1971). GD, Granite(Fookes *et al* 1971). RB, Basalt, Robertson, New South Wales, Australia. DP, Dolomitic phyllite, Broken Hill, New South Wales, Australia. CP, Calcareous phyllite, Broken Hill, New South Wales, Australia.

4.5.3 ROCK QUALITY DESIGNATION

Since RQD is a function of fracture spacing and of the condition of fractures, it would be expected that it would be a measure of the degree of weathering. Very little data are available, but it is suggested (Fig. 4.16) that RQD decreases as

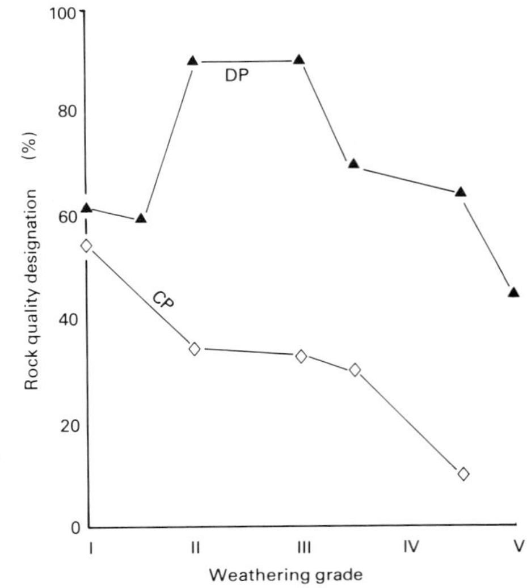

Fig. 4.16 Variation in RQD with weathering. DP, Dolomitic phyllite, Fowlers Gap, New South Wales, Australia. CP, Calcareous phyllite, Fowlers Gap, New South Wales, Australia.

weathering grade increases from grade III. For rock less weathered than grade III, RQD is dependent more on fracture spacing alone, and is independent of weathering, so it is not a sensitive indicator of weathering in grades I and II.

4.5.4 ROCK:SOIL RATIO

It has been recognized by engineering geologists working in the field that the ratio of rock to soil in a weathered mass of crystalline rock is a critical index of weathering grade. It is possible, in the field, to determine this ratio quite simply: it is equally simple to determine in core logging. The ratio is an essential index in

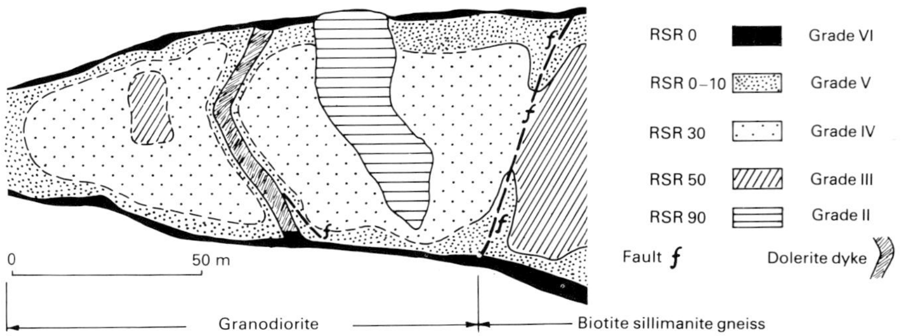

Fig. 4.17 Rock soil ratio map for part of Rocky Valley Dam foundation trench, Bogong High Plains, Australia.

Dearman's classification of weathered granitic rocks, especially in the separation of grades III, IV, and V. It is not always applicable in sedimentary rock masses. Figure 4.17 shows the rock : soil ratio map for a gneiss–granodiorite mass at Rocky Valley Dam, Kiewa, Australia. Table 4.10 shows the rock : soil ratio distribution used in mapping the dam foundations.

Table 4.10 Rock : soil ratio for weathered crystalline rocks.

Weathering grade	RSR	Remarks
Fresh I	95–100	Some soil on major joints
Slightly weathered II	90–95	Some soil on all joints
Moderately weathered	60–90	Advanced weathering on all joints. Core stones large
Highly weathered IV	30–60	Small core stones in soil mass
Completely weathered V	0–30	Rare small core stones

4.5.5 WEATHERING MICRO-INDEXES

Indexes such as fracture spacing, point load strength, RQD and RSR are determined on the material, and the mass, in the field. It has been proposed recently that micro-indexes, which are determined in the laboratory on the rock material examined microscopically, are of considerable value. These indexes are, at present, of interest more to research than to field workers, but they can be used to supplement and refine field classification. The indexes are intended to assist in grade classification by close examination of the rock material. The two indexes are I_{mp} the micropetrographic index and: I_{fr}, the microfracture index. The micropetrographic index is defined as:

$$I_{mp} = \frac{\text{percentage sound constituents}}{\text{percentage unsound constituents.}}$$

The 'sound constituents' are the primary minerals, while 'unsound constituents' are secondary minerals, produced by weathering, such as sericite, clays and iron oxides, together with microcracks and microvoids resulting from weathering. The microfracture index is the count of the number of microcracks in a 1 cm traverse of thin section. The microcracks counted include stained grain boundaries, open grain boundaries, stained microcracks in quartz and feldspar, infilled microcracks in quartz and feldspar, and clear transgranular microcracks crossing grains and pores in plagioclase. Table 4.11. shows the range of values of these indexes for a weathered granite.

It is apparent that these indexes could be determined readily only for coarser textured rocks: application to the finer grained sedimentary and metamorphic rocks would be difficult.

4.6 The effects of weathering on geomechanical properties

Weathering processes result in a general deterioration in the properties of the

Table 4.11 Micro-indexes of weathering for a granitic rock. [After Irfan and Dearman (1978).]

Weathering grade	I_{mp}	I_{fr}
Fresh I	>12	<0.5
Slightly weathered II	6–12	0.5–2
Moderately weathered III	4–6	2–5
Highly weathered IV	2–4	5–10
Completely weathered V	<2	>10

rock mass and the rock material. The reasons for this are obvious: weathering results in the weakening and breaking of bonds between mineral grains in the material, and between blocks in the mass. Minerals with high strength and elasticity are altered to minerals with low strength and elasticity, while significant changes in fabric patterns occur. In trying to assess the effect of weathering, it will be found that little systematic data are available, but some work on granites, clastic sediments, and low grade metamorphic rocks is sufficient to draw some tentative conclusions relating to compressive and tensile strengths, elasticity, density, moisture content, porosity and permeability.

4.6.1 UNIAXIAL COMPRESSIVE STRENGTH

The onset of weathering results in an immediate, and often considerable, reduction in the compressive strengths of rocks (Fig. 4.18, Table 4.12). In rocks of

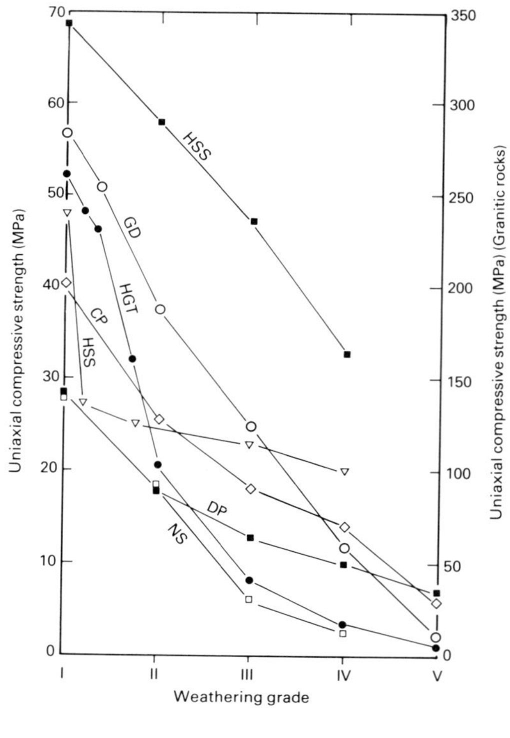

Fig. 4.18 Variation in uniaxial compressive strength with weathering. HSS, Hawkesbury Sandstone, Sydney, New South Wales, Australia. GD, Granite (Baynes *et al* 1978). HG, Harcourt Granodiorite, Victoria, Australia. CP, Calcareous phyllite, Broken Hill, New South Wales, Australia. DP, Dolomitic phyllite, Broken Hill, New South Wales, Australia. NS, Narrabeen Sandstone, Sydney, New South Wales, Australia.

Table 4.12 Uniaxial compressive strength of weathered rocks.

Rock	Uniaxial compressive strength at weathering grade shown (MPa)								
	I	II	IIi	IIii	IIiii	IIiv	III	IV	V
Hingston Down Granite	282			251		187	123	58	11
Harcourt Granodiorite	260		240	230	160	102	40	18	5
Hawkesbury Sandstone	48		27		25		23	20	
Hawkesbury Sandstone	69	58					47	33	
Narabeen Sandstone		18					6	2.5	
Narabeen Claystone		14					12		
Dolomitic Phyllite	28	18					12	10	7
Calcareous Phyllite	40	25					18	14	6

extremely high strength, such as granites, the loss between grades I and III may be as high as 80%. Even in lower strength rocks, the loss may be up to 60%. When weathering proceeds beyond grade III, strength loss continues, but is relatively much less. From this it may be inferred that the weakening, and breaking, of bonds between mineral grains, and the development of microfractures, all of which are characteristic of the transition from grade II to grade III are responsible for the loss of strength. At this stage also, the replacement of strong minerals by the weak clay minerals begins to be effective.

The loss of the strength in the rock mass is almost certainly greater than that in the material, since tests on the material at grades III and IV are made on core stones, and the mass at these grades consists of a large amount of soil. The soil, relative to the rock material, has little or no strength.

4.6.2 TENSILE STRENGTH

Virtually no published data exist relating to the effect of weathering on tensile strength. Tensile strength of rock is invariably low: the few test results available, all for clastic sedimentary rocks, suggest that the effects of weathering on tensile strength are not always as dramatic as is the case for uniaxial compressive strength (Table 4.13, Fig. 4.19). It will be noted that, for the Hawkesbury

Table 4.13 Tensile strength of weathered rocks.

Rock	Tensile strength (MPa)			
Weathering zone	I	II	III	IV
Hawkesbury Sandstone	4.15	3.30	3.56	n.d.
	8.36	6.48	4.27	3.11
	4.82	4.16	3.91	3.83
Bulgo Sandstone	4.58	1.38	0.74	0.42
		0.68	0.44	0.38
Wombarra Claystone	6.32	1.61	0.30	0.02
	3.51	0.78	0.13	0.00

Sandstone, loss of tensile strength is gradual; in the case of the Bulgo Sandstone, a 70% loss of strength occurs from grade I to grade II. This is due to the quite different textures and fabrics of the two sandstones. The 75% loss of tensile strength in the claystone between grades I and II is due to the very strong development of fractures, with even slight weathering, in this rock.

4.6.3 ELASTICITY

As would be expected, the increase in fracturing which accompanies weathering, and the replacement of brittle minerals by soft clays, result in a significant loss of elasticity with weathering (Table 4.14, Fig. 4.20).

With inadequate data, it is not possible to develop broad, general statements relating to the effects of weathering on elasticity. Values for Poisson's ratio are

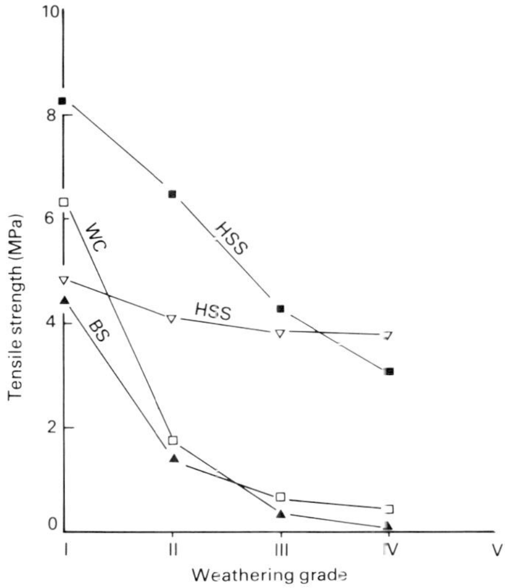

Fig. 4.19 Variation in tensile strength with weathering. HSS, Hawkesbury Sandstone, Sydney, New South Wales, Australia. WC, Wombarra Claystone, Wollongong, New South Wales, Australia. BS, Bulgo Sandstone, Wollongong, New South Wales, Australia.

Table 4.14 Tangent modulus of elasticity of weathered rocks.

Rock	Tangent modulus of elasticity (GPa)								
	I	II	IIi	IIii	IIiii	IIiv	III	IV	V
Granite	69.5		57.6		40.2	32.5	22.8	12.7	1.5
Sandstone	10.7		7.8		7.6		5.5	4.6	
Sandstone	24.6	18.9					10.7	9.2	
Sandstone	11.0	9.0					8.7	7.5	
Claystone		3.9					1.9		
Phyllite	4.9	2.6					1.9	1.6	
Phyllite	2.5	2.1					1.6	1.4	0.8
Marble	4.2	2.6					1.9		

presented in Table 4.15. In this case, virtually no data are available on a systematic basis and, as can be seen, this parameter may decrease in value with increased weathering, or it may increase.

As the test results apply to the rock material, it would be expected that changes in elasticity for the rock mass would be more severe. This is reflected in the change in ultrasonic velocities in the rock mass with weathering, since these velocities are a function of elasticity. This fact led Iliev (1967) to propose the following coefficient of weathering:

$$k = V_0 - (V_w / V_0)$$

where V_0 is the ultrasonic velocity in fresh rock, and V_w that in weathered rock. Using this coefficient permits some assessment of the elasticity of the rock mass.

4.6.4 MOISTURE CONTENT, POROSITY AND DENSITY

As weathering progresses, the general tendency is for moisture content and porosity to increase, and for density to decrease. The changes in density and moisture content are a direct result of the increasing porosity, which itself is due

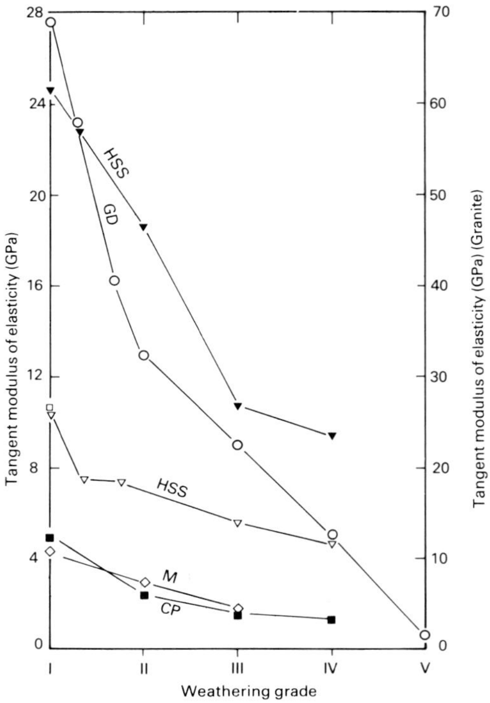

Fig. 4.20 Variation in tangent modulus of elasticity, at 50% ultimate strength, with weathering. GD, Granite (Baynes *et al* 1978). HSS, Hawkesbury Sandstone, Gosford, New South Wales, Australia. CP, Calcareous phyllite, Broken Hill, New South Wales, Australia. M, Marble, Angaston, South Australia.

Table 4.15 Poisson's ratio of weathered rocks.

Rock	I	II	III	IV
Granite	0.25	0.23	0.35	
Granite	0.25	0.19	0.19	0.27
Sandstone	0.22	0.35	0.17	
Claystone	0.36	0.33	0.42	
Sandstone	0.13	0.18	0.23	0.23

to solution and to an increase in fracturing which accompany weathering. Trends for some rocks are presented in Tables 4.16, 4.17 and 4.18 and in Figs 4.21, 4.22 and 4.23.

Table 4.16 Porosity of weathered rocks.

Rock	Porosity (%)				
	I	II	III	IV	V
Sandstone	14.18	20.70	21.52	22.07	
Sandstone	9.03	10.55	12.46	14.32	
Granite	0.27	2.78	3.78	9.60	20.20
Slate	5.38	11.00	17.65	15.40	18.77
Phyllite	7.08	8.11	11.83	10.01	

Table 4.17 Density of weathered rocks: saturated and dry.

	Density (Mg m^{-3})									
Rock	I		II		III		IV		V	
	SD	DD	SD	DD	SD	DD	SD	DD	SD	DD
Sandstone		2.22		2.09		2.05		2.03		
Sandstone	2.56		2.40		2.30		2.18		2.00	
Slate	2.74	2.69	2.40	2.29	2.42	2.19	2.46	2.31	2.02	1.85
Granite	2.61		2.58		2.56		2.43		2.23	
Phyllite	2.69	2.62	2.63	2.55	2.53	2.42	2.55	2.45		

Table 4.18 Saturation moisture content of weathered rocks.

	Saturation moisture content (%)				
Rock	I	II	III	IV	V
Sandstone	2.10	2.60	3.52	5.06	7.41
Slate	2.00	4.84	7.97	6.67	10.16
Granite	0.11	1.09	1.52	4.13	10.02
Phyllite	2.70	3.18	4.90	4.22	

The changes noted for rock material are even more effective in the rock mass. The opening of fractures in the mass results in a dramatic increase in 'porosity' of the mass, with related changes in density and moisture content.

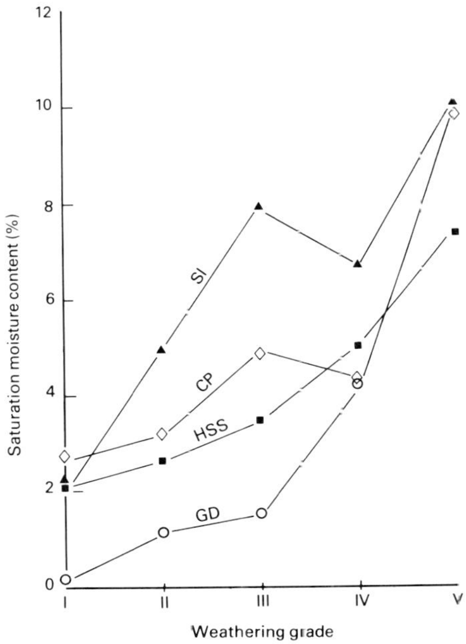

Fig. 4.21 Variation in saturated moisture content with weathering. GD, Granite (Baynes *et al* 1978). CP, Calcareous phyllite, Fowlers Gap, New South Wales, Australia. Sl, Slate, Mudgee, New South Wales, Australia. HSS Hawkesbury Sandstone, Gosford, New South Wales, Australia. HSS Hawkesbury Sandstone, Gosford, New South Wales, Australia.

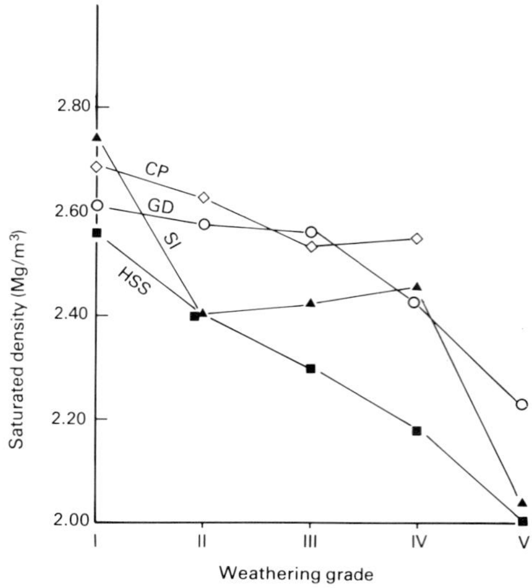

Fig. 4.22 Variation in saturated density with weathering. GD, Granite (Baynes *et al* 1978). CP, Calcareous phyllite, Fowlers Gap, New South Wales, Australia. Sl, Slate, Mudgee, New South Wales, Australia. HSS, Hawkesbury Sandstone, Gosford, New South Wales, Australia.

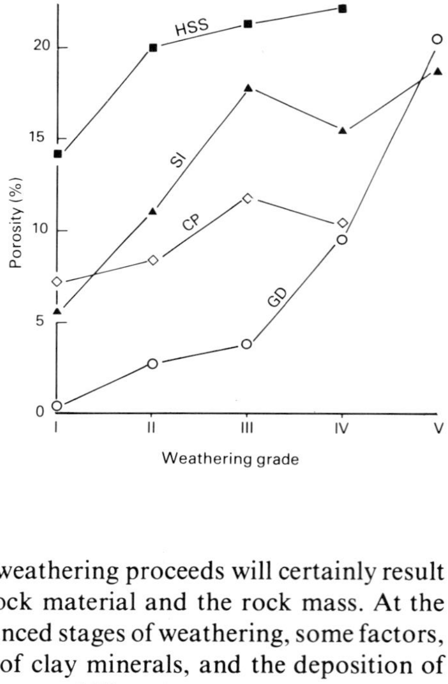

Fig. 4.23 Variation in apparent porosity with weathering. HSS, Hawkesbury Sandstone, Gosford, New South Wales, Australia. Sl, Slate, Mudgee, New South Wales, Australia. CP, Calcareous phyllite, Fowlers Gap, New South Wales, Australia. Granite (Baynes *et al* 1978).

4.6.5 PERMEABILITY

The changes in porosity which occur as weathering proceeds will certainly result in changes to the permeability of the rock material and the rock mass. At the same time, particularly in the more advanced stages of weathering, some factors, such as the development and dispersal of clay minerals, and the deposition of secondary cements, may act to reduce permeability.

Changes in primary permeabilities of granite from 10^{-6} to 10^{-8} cm s^{-1} in slightly weathered granite, to 10^{-4} cm s^{-1} in completely weathered rock have been recorded. Measurements of permeabilities of phyllites have shown an increase from 6×10^{-9} cm s^{-1} for the fresh rock to 3×10^{-8} cm s^{-1} for completely weathered rock.

The rock mass permeability (mainly a secondary permeability) certainly increases with weathering, but again, the filling of fractures by debris may exert an inhibiting influence on the flow of water through the mass. No published data which permit a systematic study of the permeability of a rock mass appear to exist.

4.7 Weatherability and durability of rock

Rock is likely to degrade while forming part of an engineering structure: the durability of the rock material in this context is defined as the resistance of the rock to short term weathering processes; it is the inverse of weatherability. Weatherability is a measure of the amount of weathering which will occur in a given time, for example, for the full life of an engineering structure. A great many tests have been devised to measure both weatherability and durability, and all are of dubious value, since no single test can hope to simulate all of the factors which operate in nature.

A test has been devised to assess the *slake durability* of clastic sedimentary rocks with a high proportion of clay-size particles, and of other rocks, the weathering of which has produced clay and other secondary minerals. The test is ideal for defining the difference between a soil or rock. Unfortunately, both geologists and engineers have misused the test, and misinterpreted test results. The literature not infrequently contains references to the slake durability index of fresh igneous rocks! The test measures the durability of a clastic sediment, or of a weathered rock, with respect to alternate wetting and drying, and to abrasion. It is of dubious value as an index of weathering (Table 4.19) or even as a predictor of durability or weatherability.

Table 4.19 Slake durability indexes for weathered rocks.

	Slake durability index I_D (%)				
Rock	I	II	III	IV	V
Phyllite	84	86	88	71	71
Claystone	97	n.d.	75	n.d.	n.d.
Sandstone	83	n.d.	71	68	n.d.
Sandstone	98	96	83	61	n.d.
Sandstone	95	94	90	87	n.d.
Sandstone	90	89	92	98	52

Several hundreds of slake durability tests have been carried out systematically on sandstones and shales from the Sydney Basin. It was found that, due to secondary silicification in some highly weathered rocks, I_D for these was much higher than for fresh and slightly weathered rocks. Moreover, I_D values for sandstones, claystones and shales were frequently comparable, although field observation clearly indicated that in both long term and short term weathering the sandstones were much more durable than the finer clay rocks.

Experience is suggesting that slake durability should be used only as a classification test, and that even then the results of tests must be interpreted in terms of the petrography of the rock. In particular, the test does not incorporate any aspect of decomposition, as distinct from physical disintegration.

Any reliable test must include both physical and chemical aspects of weathering.

4.8 Weathering of biogenic rocks

Weathering of limestone will depend on the composition of the rock. A pure, or almost pure, limestone will weather almost exclusively by solution which begins along discontinuities, and rapidly develops into large cavernous openings, in which deposits of secondary calcite may occur. If the limestone is impure (Fig. 4.24), solution remains one of the main weathering processes, but other chemical processes will operate, depending on the nature of the impurities. In coastal regions, solution of limestone will be inhibited.

Fig. 4.24 Solution weathering of impure siliceous dune limestone, in coastal cliffs, Gunnamatta, Victoria, Australia.

In a rock from southern England, which is a contact metamorphosed siliceous limestone, fracture spacing is more or less constant at 0.15 m through all grades, but decreases to 0.03 m in grade V. Point load strength varies, with grades III – IV rock having the highest value. Porosity is low in grade II (0.55%) and grades III and IV (1.1 – 1.35%), increasing dramatically to 58.5% in grade V.

A Palaeozoic limestone at Lilydale, Australia, shows solution openings along joints and bedding planes, and on some of these fractures, secondary clay minerals occur derived from silicate impurities in the limestone. Large solution cavities have been filled with surface debris. Grade IV rock has fracture spacing of 0.05 m; grade II, 0.1 m and grade I, 0.15 m.

A strongly metamorphosed limestone north of Broken Hill, Australia, shows virtually no sign of weathering beyond small solution channelling on the exposed surfaces, and some slight solution of joints, infilled by surface material. The mass as exposed is grade II; the material, grade I. In limestones which have a relatively high degree of purity, weathering seems normally to be expressed by increased fracturing.

Weathering of coal, in engineering terms, has been examined only very rarely. One example is the weathering of a brown coal from Yallourn, Australia, in which the onset of weathering is marked by the development of a soft, highly compressible, organic clay, along discontinuities. Complete weathering has produced a black, organic, highly thixotropic and highly compressible clay up to 1 m thick which terminates sharply on fresh coal. It was found that the completely weathered coal had quite inferior engineering properties, while the presence of this material on joints in the coal mass resulted in a significant decrease in the shearing resistance along the joints, and to unexpectedly high settlements under foundation loading.

4.9 Weathering in engineering practice

The nature, degree, and pattern of rock weathering are commonly among the most important of the geological factors to be considered at an engineering site. The design and construction of foundations, tunnelling methods, slope stability and the availability of construction materials may all be influenced by the weathering of both the rock material and of the rock mass. In the engineering geological investigation of a site, considerable attention has to be given, therefore, to the careful determination of weathering characteristics. Reliable weathering maps and cross sections based on direct surface observations, drilling, and geophysical investigations should be prepared In addition, statements describing the effects of weathering on engineering properties of the rocks, and on fabric and mineralogical changes which influence these properties, should also be prepared. It is essential, also, that observations be made during the construction phase of the project, when not only can checks be made on the investigation, but also additional data may require modifications to be made to the design.

4.9.1 ROCK WEATHERING AND DAM FOUNDATIONS

For the most part, concrete dams must have their foundations located in essentially fresh rock. For some, such as mass gravity and buttress types, some slight weathering of the rock mass, e.g. staining of fractures, may be permissible. For arch dams, however, both rock material and rock mass are generally required to be fresh. Earth dams on rock foundations are almost invariably constructed on weathered rock, not uncommonly in the highly to completely weathered condition, although some conservative designers prefer a foundation in the moderately weathered condition. Even in the case of foundations for earth dams, however, knowledge of the weathering is required to plan and design such aspects as concrete cut-off walls and grouting programmes.

Clover Dam, on the Kiewa River, Australia, is a small concrete buttress dam founded on a coarsely crystalline, strongly jointed and weakly foliated sillimanite biotite gneiss which is deeply weathered. Faulting has occurred at the site, and a large number of lamprophyre dykes occur intrusive into the gneiss (Fig. 4.25).

The pattern of weathering is asymmetric, with the fresh rock surface on the east abutment much deeper than on the west. Designers required excavation down to essentially fresh rock. The lamprophyre dykes were weathered to a somewhat greater depth than the gneiss, so that where a buttress was to be founded on lamprophyre, even deeper excavation was required.

The most important aspect of the weathering at this site was the development of sheeting. These structures, sub-parallel to the surface, occurred at vertical

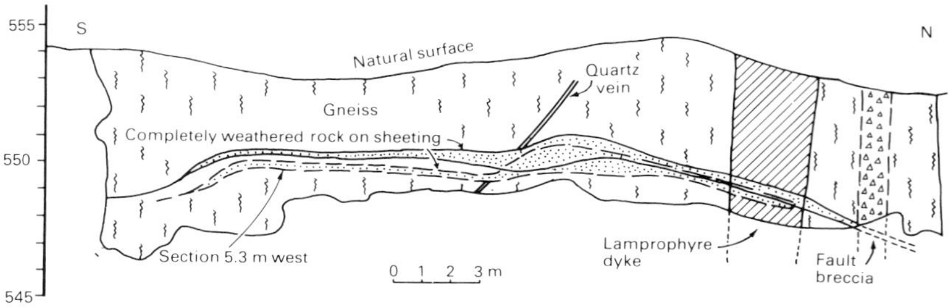

Fig. 4.25 Section along the centre line of a buttress foundation of Clover Dam showing fault zone, lamprophyre dyke, and sheeting structure.

intervals of about 5 – 10 m, and each consisted of from 10 to 25 cm of completely weathered gneiss. Drilling showed that these structures persisted to below design foundation level. During excavation, slope instability developed, with sliding along the sheeting, causing construction difficulties and delays. Moreover, during excavation, the unloading tended to accelerate weathering on the sheeting, so that, on two of these structures, the thickness of weathered material increased at a steady rate of 1 cm per month. Finally, these structures constituted a seepage problem around the abutments of the dam. Acceptance of portland cement grout was low, and careful flushing was required before the sheeting could be effectively grouted.

Similar problems with sheeting due to weathering have been reported from the Hiwassee Dam in North Carolina, USA. Here, the sheeting structures, each a few centimetres to 0.7 m in thickness, occurred at intervals of about 2 m to a depth of at least 20 m below the river bed. In this case, excavation in the channel section was carried down to about 18 m below river bed level.

Limestone is a highly soluble rock, and, in the weathering process, solution cavities, which may be open, or partly to completely filled with rubble, are developed. Solution normally begins along joints and bedding planes; in some cases, as at the Fontana Dam on the Tennessee River, USA, the solution has proceeded no further. In other cases, extensive solution has produced extensive large cavities, as at the Kentucky Dam, Tennessee River, USA (Fig. 4.26), where solution has extended to over 50 m below the surface.

These weathering characteristics of limestone have two major effects at a dam site. Probably the more important is the question of leakage from the storage. The Monte Jacques Dam, a concrete arch structure at Andalusia, in Spain, was constructed on cavernous limestone, and, after the initial filling, the storage drained through limestone caverns. The storage has never held water since, despite strong attempts at carrying out remedial measures. In the case of the Kentucky Dam, remedial measures using hay, bitumen and cement grout were successful, but the cost of these measures was extremely high.

The second, but somewhat lesser, problem associated with cavernously weathered limestones and dolomites concerns the bearing capacity of the foundation rocks. Solution openings, whether or not rubble filled, have the effect of reducing the strength of the rock mass.

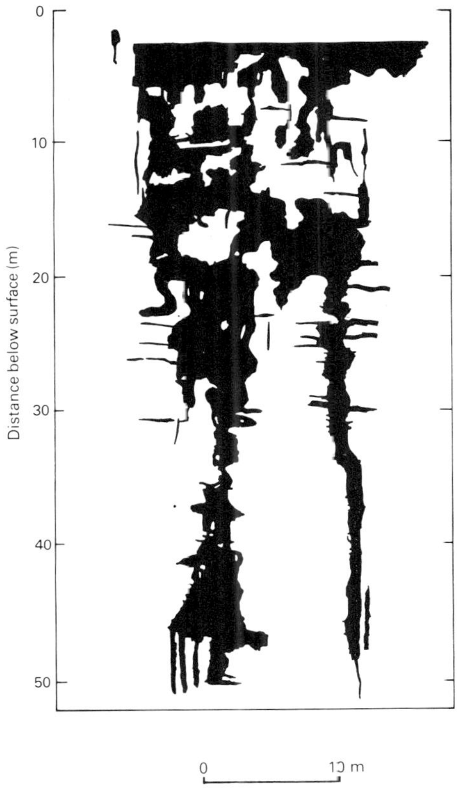

Fig. 4.26 Solution in limestone at Kentucky Dam, USA. [After Burwell and Moneymaker (1950).]

Remedial measures against leakage may also be satisfactory, however, in improving the other engineering characteristics of the mass.

In arid regions solution problems may occur after dams are in service. Under arid climatic conditions, soluble salts such as halite and gypsum may be deposited near the surface as part of the weathering process. When a water storage is in service, seepage from the storage may result in solution of these salts, producing a quite dramatic increase in seepage losses. The presence of these salts can effect the engineering characteristics of the soil in which they occur. Soils so affected require special consideration during design and construction.

4.9.2 ROCK WEATHERING IN RELATION TO UNDERGROUND EXCAVATIONS

In tunnel and underground power station design and construction, the excavation may be in soft rock or in hard rock; the latter is usually preferred for underground hydroelectric development. In the case of hard rock, excavation methods are usually planned on the assumption of fresh rock conditions. Similarly, for stability purposes, designers normally require that the machine hall and pressure tunnels, at least, be located in fresh rock. Deep and intense weathering creates special difficulties in portalling operations, especially as portals are frequently located in steep slopes, the stability of which may be marginal.

Where the rock is in a highly or completely weathered condition, normal hard rock tunnelling methods are not economically applicable, and moreover, very strong supports will be required.

In the case of pressure tunnels and hydroelectric underground machine halls, design engineers usually prefer a fresh rock cover of thickness varying between 20 and 50 m. The thickness appears frequently to be chosen arbitrarily, with no regard for the *degree* of weathering, nor for the mechanical properties of the weathered rock.

Weathering was one very important geological factor in the location and design of the McKay Creek hydroelectric power station and tail race tunnel, in Australia. At the site, in a faulted and jointed granodiorite, weathering was complete, for a considerable depth, while that associated with fault zones was even deeper. Seismic refraction surveys and diamond drilling were used to define the fresh rock surface and to locate fault zones, and it was on the basis of the results of these investigations that the final site for the power station was selected (Fig. 4.27). Even so, the cover of fresh rock at the south-western end of the power

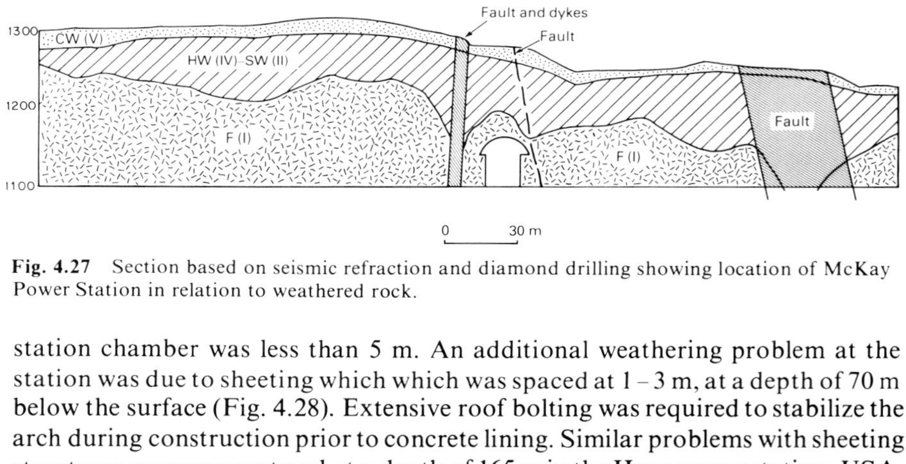

Fig. 4.27 Section based on seismic refraction and diamond drilling showing location of McKay Power Station in relation to weathered rock.

station chamber was less than 5 m. An additional weathering problem at the station was due to sheeting which which was spaced at 1 – 3 m, at a depth of 70 m below the surface (Fig. 4.28). Extensive roof bolting was required to stabilize the arch during construction prior to concrete lining. Similar problems with sheeting structures were encountered at a depth of 165 m in the Haas power station, USA.

At the investigation stage, some interesting problems were anticipated at the Jor station, Malaysia, where tropical weathering extended to a depth of 70 m, and where the fresh porphyritic granite showed a marked tendency to very rapid physical disintegration on exposure. However, during construction at a depth of 260 m, anticipated problems did not arise. On the other hand, at Turlough Hill, Ireland, the coarse textured granite is completely weathered at the surface, and weathering effects extend down to station level, 56 m below the surface. Although the rock material is fresh, the rock mass is to be regarded as moderately weathered. Joints at station level were infilled with weathered material — illite and montmorillonite. Flat sheeting structures were present, and, as a result, it was decided to design the upper part of the station with an

Fig. 4.28 McKay Underground Power Station under construction. Note sheeting on left hand wall.

elliptical cross section to achieve stability. In addition, prestressed rock anchors were used to increase stability.

The Ramu underground power station in New Guinea was constructed in marble, intruded by thin dolerite sills. Although investigation indicated the rock at the site was fresh, excavation encountered debris-filled solution cavities. One of these occurred at roof level at the south end of the roof, and caused considerable problems with stability.

The Liapootah Tunnel, Tasmania, was driven in a thick dolerite sill, in which advanced weathering extends for up to 100 m below the surface. The base of the weathered zone is extremely irregular, and a number of sections of completely weathered rock were encountered in the tunnel. These sections were wet and clayey, and occasionally, open cavities occurred. Heavy steel ribs were required as support in these sections pending the placing of reinforced concrete lining.

In Bavaria, the Reisach and Weinberg pressure tunnels were constructed in moderately to highly weathered gneiss, at depths with cover often less than 50 m. Here, the designers adopted a rule of cover equal to half the internal hydrostatic pressure, which, for the time, was quite radical. However, thick lining and a careful grouting programme resulted in stable structures. Considerable effort was expended in flushing clay size material from the weathered rock about the tunnels prior to grouting.

4.9.3 WEATHERING AND SLOPE STABILITY

The weathering of rock constituting a slope can reduce the stability of the slope with respect to sliding in several ways. The most obvious is the production of soil which is more prone to sliding, than rock. Another way in which stability is reduced is by the opening up of structural defects, which then not only constitute potential surfaces of sliding, and reduce the strength of the rock mass, but also allow the admission of water, which increases the stresses on the slope. A third type of instability is produced by differential weathering, and this is of particular importance on cliffs developed in sedimentary rocks. The more weatherable rocks decompose, disintegrate and are eroded more rapidly than the more weather-resistant rock, often resulting in sections of the slope being unsupported. The point can be reached when the stresses produced in this way exceed the shear strength of the rock, and failure occurs.

During the construction of the Clevedon Hills section of the M5 motorway in England, weathering of limestone and sandstone created stability problems in the constructed batters. In the upper 2–3 m, the rocks were in the slightly to moderately weathered condition, with considerable fragmentation. Retaining walls were required in the upper section of the batters to ensure stability. Below about 3 m, the open joints due to weathering required bolts and anchors, and locally, masonry walls were constructed (Fig. 4.29).

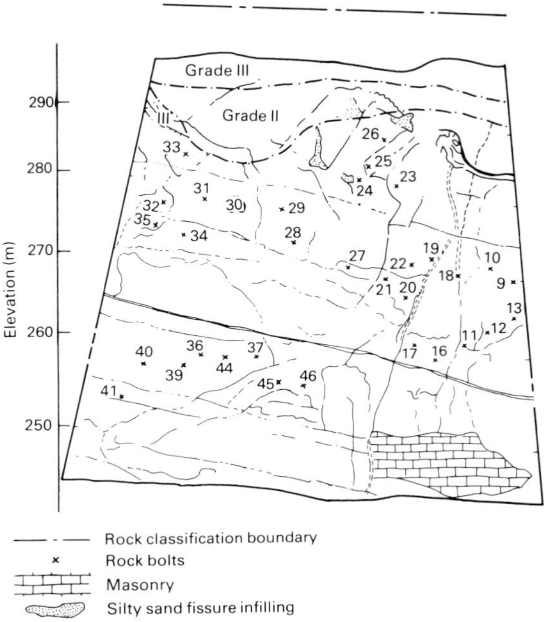

Fig. 4.29 Weathering pattern and remedial works, M5 Motorway, England. [After Eyre (1973).]

Along the Illawarra Escarpment, south of Sydney, Australia, large rock falls, and rock slides, occur relatively frequently, endangering roads and residential areas constructed on the lower slopes. It has been shown that weathering of the massive, horizontal sandstones and thin shales is the major factor contributing to slope instability. Weathering of the shales accelerates time-dependent creep deformation, which in addition to the more rapid weathering of the shales, leads to the toppling of joint-bedding plane bounded sandstone blocks. The condition is aggravated by the opening of joints by weathering processes, which leads to a reduction in the shear strength of joints in the rock mass.

4.9.4 WEATHERING AND CONSTRUCTION MATERIALS

In terms of the availability of suitable materials for construction, weathering can be regarded as both beneficial and deleterious. Earth construction materials are the result of weathering processes. In the case of residual soils, the quantities available will depend on the degree and depth of weathering of the parent rock, and the weathering processes can influence the mineralogy of the soils, which in turn will affect the engineering properties.

Rock for use as rip-rap, or as concrete or road aggregate, is required, generally, to be in a fresh condition. Any effects of weathering can have quite harmful results. The presence of certain clay minerals, for example, can produce damage to concrete, and opaline material in slightly weathered rocks can react with high alkali portland cements to produce expansion and disruption of concrete. Weathered surfaces on particles of natural gravel result in a poor bond with cement matrix.

In the construction of the main runway of Sydney Airport, very slightly weathered dolerite was used as rip-rap in that section constructed in Botany Bay. Reaction between clay minerals, already present, and sea water, resulted in very rapid disintegration and decomposition of the dolerite, and so this rock had to be abandoned as a supply.

Residual soils, developed on a rock mass as a result of weathering, are a ready source of earth construction materials. Knowledge of the degree, and pattern, of weathering is essential, however, before any borrow area is developed. Moreover, the soil itself can reflect minor changes due to weathering. In a granitic soil used for the construction of the Rocky Valley Dam, Australia, it was found that within the completely weathered rock, textural changes occurred reflecting a slight decrease in degree of weathering with depth. At, and near the surface, the texture was much finer, due to advanced granular disintegration, each soil particle being monomineralic. With increasing depth, and with weathering only very slightly less advanced, the soil texture became coarser. This is due to many of the soil particles consisting of two or more mineral grains as a result of less complete granular disintegration (Fig. 4.30).

4.10 Concluding remarks

Rock weathering is of significance to the engineer, and the engineering geologist, in almost every aspect of their work. In some projects, it assumes overriding importance. It is essential for the geologist, in his investigation report, to specify the degree of weathering of rock; the pattern of weathering; the effects of weathering on engineering properties, and to develop a satisfactory classification of weathered rocks, for the particular project. This classification is critical for the engineer's interpretation of the geological information.

So far as possible, weathering data should be quantitative. This does not imply that qualitative descriptions are unnecessary. Such descriptions are essential, but they are not, in themselves, sufficient.

The possible effects of weathering of the rock on the engineering work should be stated. Weathering affects strength, elasticity, settlement characteristics, stability, durability, and, in terms of materials, utility. A considered opinion, when relevant, relating to the excavatability of the rock, should be given.

It must always be remembered that rock type, homogeneity of the rock mass, climate and landform all influence the processes and products of rock

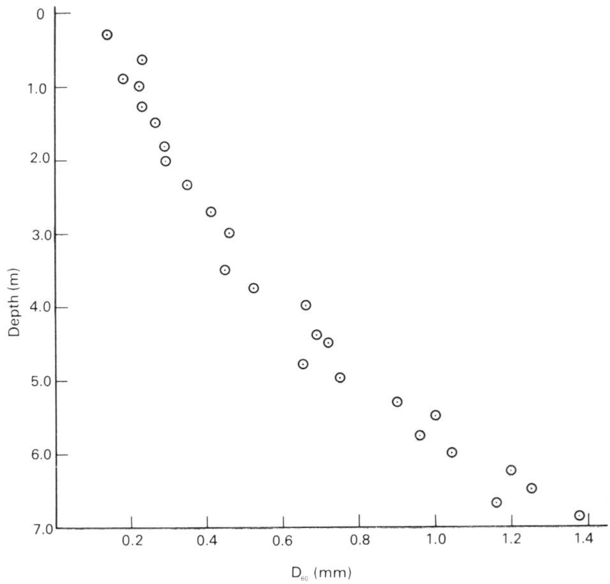

Fig. 4.30 Relationship between D60 and depth for a residual soil developed on granitic rocks, Rocky Valley, Bogong High Plains, Victoria, Australia.

weathering. Because of this, each site area or region must be regarded, in a sense, as unique, and studied carefully, applying general principles.

References and further reading

Anon. (1977) The description of rock masses for engineering purposes. *Quart. Jl. Engng. Geol.* **10,** 355–88.

Baynes F. J. & Dearman W. R. (1978) The relationship between microfabric and the engineering properties of weathered granite. *Bull Int. Assoc. Engng. Geol.* **18**, 191–8.

Baynes F. J. & Dearman W. R. (1978) Scanning electron microscope studies of weathered rocks. *Bull Int. Assoc. Engng. Geol.* **18**, 199–204.

Baynes F. J. & Dearman W. R. (1978) The microfabric of a chemically weathered granite. *Bull Int. Assoc. Engng. Geol.* **18**, 91–100.

Baynes F. J. & Dearman W. R. & Irfan T. Y. (1978) Practical assessment of grade in a weathered granite. *Bull Int. Assoc. Engng. Geol.* **18**, 101–10.

Beavis F. C. (1951) Weathering at Rocky Valley Dam, Kiewa. Internal Report, State Electricity Commission of Victoria.

Beavis F. C. (1956) Kiewa 1 Underground Power Station. Internal report, State Electricity Commission of Victoria.

Beavis F. C. (1956) The engineering geology of Yallourn D power station, Australia. *Quart. Jl. Engng. Geol.* **8**, 103–17.

Beavis F. C. (1962) Geology of the Kiewa Area. *Proc. Roy. Soc. Vict.* **75**, 349–410.

Burwell E.G. & Moneymaker B.C. (1950) Geology in dam construction. In *Application of Geology to Engineering Practice* (ed. S. Paige) pp. 11–44. Geological Society of America.

Dearman W. R. (1974) Weathering classification in the characterization of rock for engineering purposes in British practice. *Bull. Int. Assoc. Engng Geol.* **9**, 33–42.

Dearman W. R. (1976) Weathering classification in the characterization of rock: a revision. *Bull Int. Assoc. Engng Geol.* **13,** 123–8.

Evans R. S. (1978) Time dependent factors determining the rock slope stability of the Illawarra Escarpment, New South Wales. Unpublished Ph D thesis, University of New South Wales, Sydney, Australia.

Eyre W. A. (1973) The revetment of rock slopes in the Clevedon Hills for the M5 Motorway. *Quart. Jl. Engng Geol.* **6,** 223–30.

Fookes P. G., Dearman W. R. & Franklin J. A. (1971) Some engineering aspects of rock weathering. *Quart. Jl. Engng Geol.* **4,** 139–85.

Franklin J. A., Broch E. & Walton G. (1971) Logging the mechanical characters of rock. *Trans. Inst. Min. Met.* **80,** A1–9.

Franklin J. A. & Chandra R. (1972) The slake durability test. *Int. Jl. Rock Mechs Min. Sci.* **9,** 325–41.

Iliev I. G. (1967) An attempt to estimate the degree of weathering of intrusive rocks from their physico-mechanical properties. *Proc. 1st Congr. Int. Soc. Rock Mechs, Lisbon,* pp. 109–14.

Irfan T. Y. & Dearman W. R. (1978) Engineering classification and index properties of a weathered granite. *Bull Int. Assoc. Engng Geol.* **17,** 79–90.

Irfan T. Y. & Dearman W. R. (1978) The engineering petrography of a weathered granite in Cornwall, England. *Bull Int. Assoc. Engng Geol.* **11,** 233–44.

Lewis I. (1974) The stability of the coastal cliffs, Sydney, unpublished M.App.Sc. thesis, University of New South Wales, Sydney Australia.

Loughnan F. C. (1969) *Chemical Weathering of Silicate Minerals.* Elsevier, New York.

Moye D. G. (1955) Engineering geology for the Snowy Mountains Scheme. *Jl Inst. Engrs Australia* **27,** 281–99.

Ollier C. D. *Weathering.* Oliver & Boyd, Edinburgh.

Ruxton B. P. & Berry L. G. (1957) Weathering of granite and associated erosional features in Hong Kong. *Bull Geol. Soc. America* **68,** 1263.

Thomson P. J. (1967) Geological investigations for the Wyangala Spillway Site. *Civ. Eng. Trans. Inst. Engrs Australia* **CE9,** 1–44.

5

The Land Surface

5.1 Introductory statement

Most types of landform — mountains, valleys, plains, cliffs and beaches — owe their origin to processes dependent on the action of wind and water, the latter sometimes in the form of ice. The two processes principally responsible for the forms which make up the earth's surface are erosion and deposition. Abrasive rock materials carried by wind or water, or embedded in ice are responsible for erosion; deposition occurs when the transporting agent is no longer able to carry the load of material responsible for, and produced by, erosion. This suggests that two basic types of landform can be recognized: erosional and depositional. Such a classification, however, is incomplete, since some landforms are the direct result of other processes, such as igneous activity and tectonism. Volcanic cones and domes, and lava and ash plains result from volcanic activity, while escarpments and mountains are produced by faulting and folding. Geomorphologists, therefore, classify landforms as being either constructional or destructional. Such a classification, while useful, is too simple, since constructional landforms are being modified all the time by destructional, erosive, processes, and destructional landforms by constructional processes.

Two main factors influence landform morphology and pattern. The first of these is the climatic factor; the second is the geological factor. Climate plays a major role: it is obvious that erosion will be of a different type in each of humid temperate, arid, and arctic regions. Rock types, their relationship to each other, and their structure are of at least equal importance with climate. Landform pattern, for example, is very often determined by the geometry of geological structures. Some landforms develop irrespective of geology (for example coral reefs) and others, irrespective of climate (many coastal types). However, in most cases, both climate and geology have to be considered.

It is impossible for a geologist to assess adequately the geology of a region without taking into account landform. In fact, landform often gives the first piece of data; air photo interpretation for geological purposes is based almost exclusively on geomorphological principles. Similarly, the engineer requires a knowledge of geomorphology to assist in planning and design. The importance of geomorphology to the engineer concerned with road, river, beach, and slope engineering cannot be overemphasized. Geomorphic processes are in progress at almost every engineering site, and it is the process, as much as the product, which will concern the engineer.

5.2 Climatic control of landform

Climate influences the geomorphic and geological processes of weathering and erosion of rock both indirectly and directly; indirectly by determining the type and density of the vegetative cover, directly by determining the processes of weathering, erosion, and deposition. Climate has been used as the basis for the concept of morphogenic regions. In this concept, regions are defined in terms of annual temperature range, average annual precipitation, and terrain morphology. For each morphogenic region, the main agents and processes of erosion have been described. The concept assumes that for each region a complex system of climatic processes gives rise to typical landforms which are unique for each particular region.

Besides influencing the type of landform developed, climate determines the rate at which weathering, erosion, and deposition take place. If 'soil erosion' is excluded, rates of erosion are generally relatively slow. However, erosion is much more rapid in mountain areas than in lowlands, and it is slowest in arid zones.

In assessing the role of climate, it is not merely the present day climate which has to be considered. For example, vast areas of the northern hemisphere, with temperate climates, have glacial landforms developed during periods of the Pleistocene Ice Age. In the Australian arid zone, the landforms have been inherited, in part, from past humid temperate climatic periods. It is necessary, too, to consider minor seasonal climatic variations and their effects. In cold winter spells in temperate zones, frost action has a destructive effect on soil structure, which in turn has an effect on slope form and slope stability.

Coastal landforms would appear to be independent of climate (except for fjords and coral reefs). However, some coastal features such as shore platforms and foredunes appear to be climatically controlled to at least some degree.

5.3 Geological control of landform

Two elements are involved in the geological control of landform: the type of rock constituting the mass, and the structures present in the rock mass. In terms of rock type, it is necessary to consider four main groups:

(i) intrusive igneous rock masses;
(ii) volcanic rocks;
(iii) sedimentary rock masses;
(iv) metamorphic rock masses.

Structural controls are faults, folds, joints, and minor intrusive structures such as dykes. Occasionally, metamorphic foliation may be important.

5.3.1 LANDFORMS ON INTRUSIVE IGNEOUS ROCK MASSES

The topography developed on an intrusive igneous rock mass is typically rounded, with tors, slightly weathered joint bound remnants, distributed over the surface, the more highly weathered material having been eroded away. The roundness of the topography is due in a large measure to the weathering processes of exfoliation and sheeting. This does depend to some degree on rock type, since some dioritic masses tend to form topographic depressions. In some areas of very low rainfall, and where the igneous rock is very closely jointed and faulted, conical, rather than rounded, hills are formed, as in the granitic terrains of southern Sinai.

The drainage pattern in massive igneous terrains is frequently rectangular, reflecting the rectangular joint pattern of the rock mass (Fig. 5.1). In some

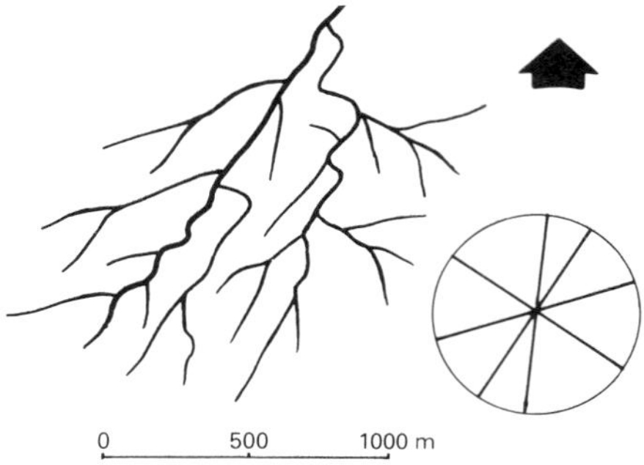

Fig. 5.1 Joint pattern and stream pattern in granodiorite, north-east Victoria, Australia.

instances, drainage lineaments may be formed (Fig. 5.2). These are drainage tracts which are rectilinear over a considerable length. Since these lineaments

(a)

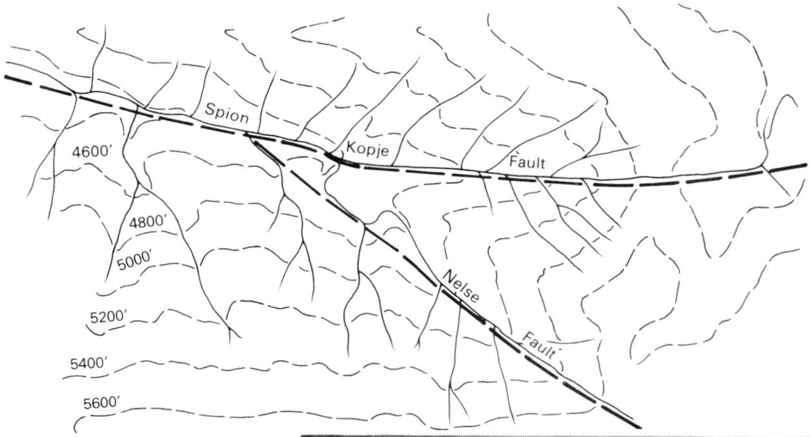

(b)

Fig. 5.2 Fault lineament on Spion Kopje Creek, north-east Victoria, Australia. (a) Map. (b) Photograph showing lineament.

represent a line of weakness comprising readily erodable material in an otherwise resistant rock mass, they are of great interest to both geologist and engineer. Drainage lineaments are most commonly formed on a fault zone, a zone of intense jointing and fracturing, or on a bed of soft rock.

5.3.2. VOLCANIC LANDFORMS

Volcanoes are points from which igneous material is erupted at the earth's surface. Eruption may occur from a single point, or from a fissure. The type of eruption, and the landforms produced depend on the nature of the molten magma. Acid and alkaline lavas are extremely viscous, and do not flow readily. As a result, they tend to form plugs, such as the needle of Mont Pelée; Shiprock, New Mexico; and the Glasshouse Mountains, Queensland, Australia. Because of the viscosity of the lavas, vents are often blocked; pressures develop, and violently explosive eruptions occur. Basic lavas, such as basalt, are more fluid, and form cones (Fig. 5.3), and lava plateaux, although violent eruptions may

Fig. 5.3 (a) Basaltic cone, New Britain. (b) Acid lava cone, New Zealand.

result in cinder cones. The Deccan, India; the Antrim Plateau of Ireland; and the Western Volcanic Plains of Victoria in Australia are examples of extensive lava plateaux.

Craters are usually present on volcanic cones. The craters are of two types: the simple crater, and the caldera. The former is the result of explosive activity, the latter of both eruption and subsidence. Both types are frequently occupied by lakes. Young lava plateaux comprise a rough terrain due not only to the broken nature of flow surfaces, but also to such features as lava blisters and lava cones. In contrast, deposits of cinders and ash (Fig. 5.4) form quite level and smooth land surfaces. After erosion step and tread topography develops (Fig. 5.5), determined by separate distinct flows. Even in this case, however, while weathering and erosion may have removed the initial roughness of the land surface, outcrops of jointed rock, and pillar strews (Fig. 5.6) result in an irregular surface, while slopes may be covered by rock flows.

Fig. 5.4 Ash deposits, Colac, Victoria, Australia.

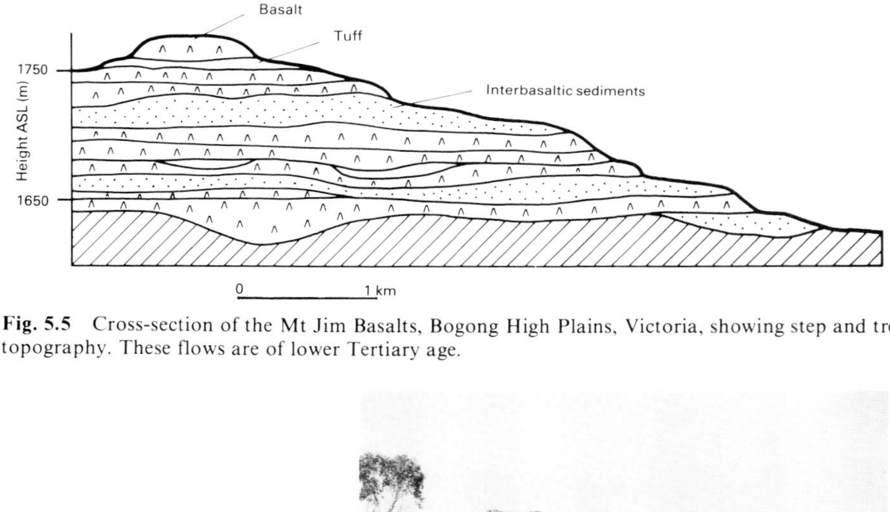

Fig. 5.5 Cross-section of the Mt Jim Basalts, Bogong High Plains, Victoria, showing step and tread topography. These flows are of lower Tertiary age.

Fig. 5.6 Jointed outcrop and pillar strews, Bogong High Plains, Victoria, Australia.

If lava flows are underlain by soft bedrock, headward erosion by streams causes a rapid undermining, collapse, and retreat of the flow margin. Consequently, waterfalls are often found at the margins (Fig. 5.7). Similar

Fig. 5.7 Waterfall at margin of basalt flow, Daylesford, Victoria, Australia.

geomorphic features may develop where lava flows are underlain by ash beds. In some cases, lava flows have partially filled pre-volcanic valleys; lateral streams develop along the margins of the flow, and, after erosion has progressed, the lava remains as an interfluvial ridge. It is important for engineers to appreciate the significance of this particular type of landform, since, at a damsite on a lateral stream, sub-volcanic sediments, with high permeability may occur, giving rise to high water losses from a storage (Fig. 5.8).

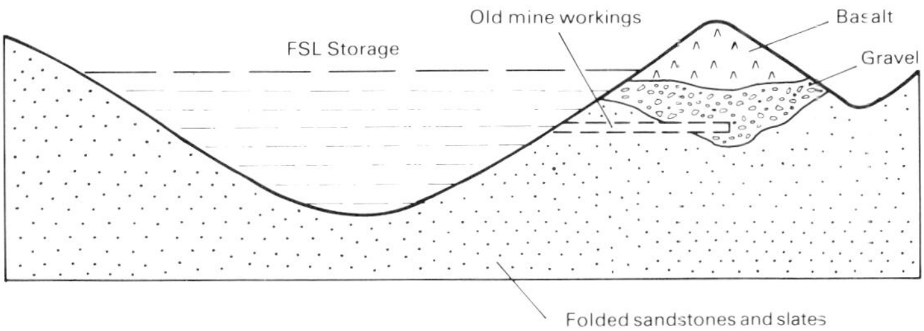

Fig. 5.8 The effect of sub-volcanic alluvial sediments at a damsite on a lateral stream.

5.3.3 LANDFORMS ON SEDIMENTARY ROCKS

Although lithology exerts some control, landform development on sedimentary rocks is more influenced by the attitude of the beds. If the sedimentary mass consists of alternating beds of hard and soft rocks, and the beds are horizontal, a dissected plateau is usually developed, in which the valleys are deep and steep sided. Near vertical cliffs are formed in the harder rocks, with somewhat gentler slopes in the soft rock. The landscape in such a terrain is frequently spectacular: examples are the Grand Canyon, Colorado, USA; and the Blue Mountains, New South Wales, Australia (Fig. 5.9). The alignment of cliffs is determined by jointing in the harder rocks, and cliff formation usually involves toppling failure

Fig. 5.9 Blue Mountains, New South Wales. A dissected plateau in horizontal sedimentary rocks.

of joint and bedding plane bounded blocks. Where dissection is not deep, step and tread topography may be developed, similar to that in volcanic terrains. When the rocks have a significant dip, but are not closely folded, a dip slope – escarpment topography is commonly formed (Fig. 5.10). In some instances the

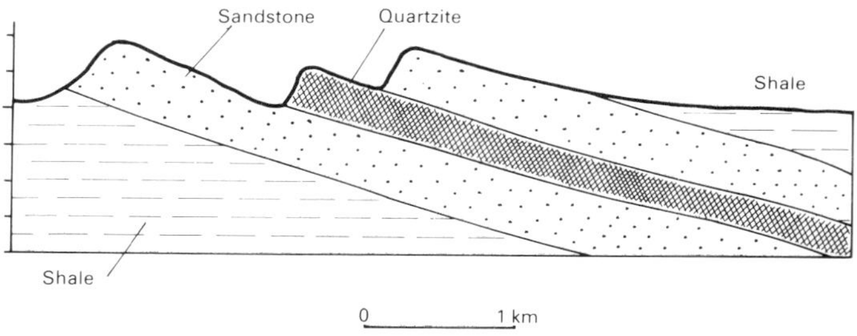

Fig. 5.10 Dip slope–escarpment topography on gently dipping sandstones and quartzites, near Broken Hill, Australia.

escarpment may be fault induced, but more commonly it is an erosive feature. Closely folded sedimentary sequences have landforms which are reflections of the fold structure. In the example shown on Fig. 5.11, hard quartzitic sandstones, on the limbs of a syncline, stand up as paired ridges, with a valley eroded into the softer shales.

Special landforms are associated with carbonate rocks such as limestones and dolomites. Typical is a karst topography, i.e. an assemblage of landforms which results from the diversion of the drainage system from the surface to

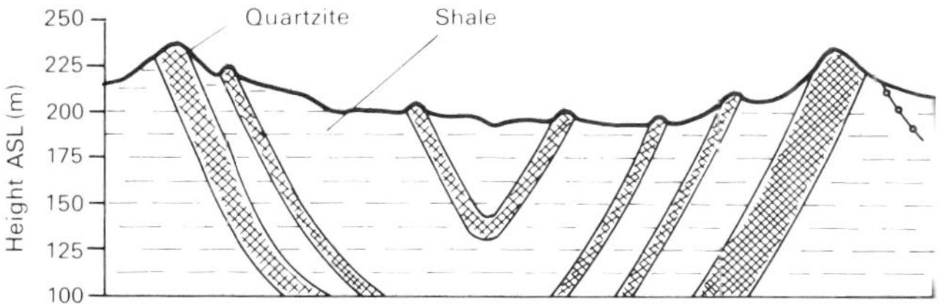

Fig. 5.11 Relationship between lithology, structure and landform, Faraway Hills, New South Wales, Australia.

underground, and with subsurface solution of the rocks. Caves and sink holes, often rubble filled, are characteristic features of such a topography. The importance of such features for the engineer requires a strong emphasis: a number of dams have been constructed, and failed to retain water, which has been lost through solution cavities in limestone, for example the Monte Jaques dam in Spain. Despite their solubility, crystalline limestones and dolomites are highly resistant to erosion, and frequently stand up as ridges and uplands.

5.3.4 LANDFORMS IN METAMORPHIC ROCKS

Because of the strong foliation in metamorphic rocks, the topography tends to have a pattern which parallels this foliation: this has been termed *textured relief* (Fig. 5.12). In some very high grade metamorphic rock masses, where the foliation is often dispersed, the rock resembles, in many respects, a granite. In such a case, the landforms are similar to those of a granitic landscape.

Fig. 5.12 Textured relief in schists with foliation dipping at a low angle, Arrowtown, New Zealand.

Contact thermal metamorphism of sediments by an intrusive igneous mass results in the conversion of the sedimentary rocks into metamorphic quartzites and hornfelses. These are extremely hard, durable, erosion resistant rocks, and, with erosion, stand up as a well defined ridge around the intrusive mass. This feature can be important for the engineer. In the 19th century, the construction of the railway from Melbourne to the Bendigo gold fields, in Victoria, involved the construction of two tunnels, one on the southern, the other on the northern, margins of the Harcourt Granite, through the ridges of hornfels.

5.3.5 LANDFORMS DUE TO FAULTS

Faults produce landforms of two basic types, depending on the fault movement, and on the pattern of faulting.

Where there is a vertical component of fault movement, an escarpment is developed. Because of the resultant uplift, erosion on the upthrown side of the fault is accelerated, and, as a consequence, most fault escarpments are quickly modified. Figure 5.13 shows an example of a deeply dissected, young fault

Fig. 5.13 Escarpment developed on the Tawonga Thrust, Mt Beauty, Victoria, Australia.

escarpment produced by a low angle thrust fault, the vertical component of movement on which was about 1000 m. It was this escarpment which provided the first evidence of a fault that was a major engineering significance in the layout design of a large hydroelectric scheme.

Paired faults may result in the production of grabens, or rift valleys, or of horsts. Some of these, for example the Great Rift Valley of Africa, constitute major terrain features.

If the fault is of a type in which the movement is essentially horizontal, i.e. a wrench fault, then no special landform may be obvious, except for the truncation of spurs and the displacement of streams. However, the crush zones of these faults are very easily eroded, so that stream valleys develop on them, to produce topographic lineaments. One of the best known of these faults is the Great Glen Fault of Scotland, marked by a deep, rectilinear valley, and a series of lochs, or lakes.

5.4 Landforms of temperate climatic zones

Water is the main agent of weathering, erosion, transportation, and deposition in temperate humid climatic regions. Hence, in such regions, streams, valleys, and other landforms produced by the action of water are dominant. Erosive and transportation activities are concentrated in streams and rivers, while deposition occurs in, and adjacent to, and at the mouths of the rivers. Every river tends to develop a profile from its source to its mouth. This profile is known as a *thalweg* (Fig. 5.14). In the source segment, the thalweg has a steep gradient, often as much as 1 in 5 or more, which decreases downstream, so that near the mouth, gradient may be 1 in 1000 or less. Often a stream encounters some resistant

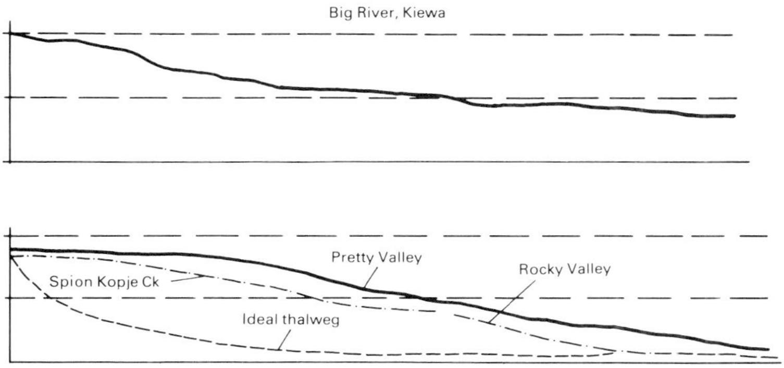

Fig. 5.14 Ideal thalweg of a stream, and thalwegs of Kiewa River tributaries, Victoria, Australia.

material, when a local base level is established, as distinct from the general base level, which is most commonly sea level, but which may be a lake. Upstream of a local base level, the gradient is flat, and the valley has often widened, with an alluviated floor. For this reason, the point at which local base level occurs may be a suitable site for a dam.

Near its source, the stream occupies a valley which is narrow and steep sided. Here, the main work of the stream is erosion and transportation. Further downstream, where gradient is reduced, the valley floor is flat and alluviated, and the stream meanders across the floor of a wide open valley. In this section, the stream is said to be graded, and both deposition and erosion take place. Finally, the stream follows, frequently, a tortuous course over a broad, open, alluvial plain. Erosion does not occur, but deposition, especially in periods of flooding, may be quite extensive. Near the mouth of the river, the sediment load may be deposited to form a delta, and the stream breaks up into a number of distributaries.

Valley formation involves both deepening and widening. The former is almost exclusively the work of the stream; the latter is partly due to lateral erosion by the stream, and partly due to mass movement of the valley walls. Mass movement includes both soil and rock creep, landslides, mudflows, block slides, and debris avalanches. Unstable valley walls are a natural feature, and even apparently stable slopes may be in a delicate state of balance which may be disturbed by any engineering activity.

5.5 Landforms of arid regions

Although arid regions have very low and irregular rainfall, and wind is the main agent of erosion, transportation and deposition, the work of water is also highly important. Rainfall, when it does occur, does so often as high intensity storms, and the resulting floods erode, transport and deposit immense volumes of material. Most arid zone streams terminate in the desert; exceptionally, a large river, such as the Nile, may cross the desert to reach the sea. In arid uplands, streams courses are rocky valleys, with steep slopes, and steep gradients. Where the streams leave the uplands, they deposit their load to form alluvial fans, which frequently coalesce to form a continuous apron. Within this apron, groundwater resources may occur. The streams here divide, and flow out onto the plains where they may die out, or terminate in an internal drainage basin.

The plains are an area of sediment accumulation, but between the uplands, and the plain, a narrow belt of bedrock, covered by a thin layer of transported rock debris may occur. This is called the pediment. Water, and the fine sediment it carries move beyond the pediment to accumulate in the centre of the basin. Any lakes formed soon evaporate leaving claypans or playas.

Apart from the work of water, wind, in arid regions, is responsible for erosion of rock due to abrasion by wind-borne sediments. The wind is also responsible for sediment movement. Deposition of sand produces dunes of various types, some of which may be fixed, others mobile. The fixed dune is a drift of sand which has accumulated against some obstacle, usually vegetation or a rocky outcrop. Mobile dunes assume many forms, but the more important are the crescent shaped barchan, the transverse dune aligned normal to the prevailing wind direction, and the longitudinal dune, parallel to this direction.

Some arid regions have had all the fine sedimentary material removed by wind action to leave the coarser rock fragments on the surface. These fragments are known as gibber, hamada, or desert armour, and the deserts are said to be stony. The stone fragments protect the underlying material from erosion: hence the term desert armour. Because of the abrasive action of wind-borne sediments, and arid zone weathering processes, many of the stones have a highly polished veneer of silica or of iron oxides; this veneer is called desert varnish. In many cases, the stones protect a very fine unsaturated clayey soil, the engineering properties of which are very different to similar soils in more humid regions.

Some arid zones, for example interior Australia, and parts of the USA, have mesa-type landforms capped by a duricrust of silcrete, clacrete or ferricrete. Often, these are associated with former alternating periods of humid and arid conditions, in Australia dating back to the Mesozoic. The duricrust cappings remain as remnants of once very extensive sheets. Beneath these cappings, the soils and rocks are generally depleted of silica, iron oxides and lime, and are often reduced in strength as a result.

5.6 Glacial landforms

Erosion and transportation in glacial regions is performed by ice; deposition occurs when the ice melts. The most common erosional landforms of glacial regions are cirques, U shaped valleys, arêtes, and fjords. Cirques, known also as corries or cwms, are extremely common in the upland areas. They are armchair shaped hollows, with steep sides, and usually have a downslope rim. Where a number of cirques have cut back into mountains, they produce sharp ridges, and pyramidal peaks called aretes. Roches moutonées are small hills produced in rock by the moving ice (Fig. 5.15). The ice polishes the gentle, long slope, and forms the steep face by rock plucking. Since the ice of glaciers truncates spurs, glacial valleys characteristically are U shaped, and the tributary valleys, also U shaped, are 'hanging', i.e. their junction is some considerable height above the floor of the main valley. The base level of ice erosion is not necessarily sea level, and may be

Fig. 5.15 Roche moutonée, Queenstown, New Zealand.

below sea level. Fjords are long, narrow, over-deepened glacial valleys, usually with a lip at the mouth, which have been invaded by the sea.

Depositional glacial landforms are common in that part of the earth's surface, and particularly in the northern hemisphere, which was affected by the Pleistocene ice ages. Two main groups of such landforms have been recognized: those due to deposition directly from ice (glacial deposits) and those due to deposition from ice melt-water (fluvioglacial deposits). Sheets of debris deposited by a glacier constitute ground moraine (Fig. 5.16). Drumlins, smooth,

Fig. 5.16 Ground moraine, near Kingston, New Zealand.

ellipsoidal hills of moraine may be associated with ground moraine. Some drumlins have a bedrock core. Terminal moraines are deposited at the front of a retreating glacier, and lateral moraines at the margins. Fluvioglacial deposits may occur as fans or outwash plains. Eskers are long, sinuous ridges of reasonably sorted, poorly stratified sand and gravel, often many kilometres long.

Under arctic or sub-arctic conditions, where precipitation is inadequate for glaciers to form, frost action results in the freezing of soil moisture to produce a layer of permanently frozen ground, called permafrost, at, or close to, the ground surface. This has particular importance for foundation engineering, due to the possibility of thawing occurring beneath a heated building, with loss of strength.

Permafrost also contributes to unstable slopes inasmuch as it forms a sliding surface for debris, produced by surface thawing, in the summer season.

5.7 Coastal landforms

Wind is an important agent in the formation of coastal landforms, since wind waves have great effect on coasts. These waves strike the shore with intense force; the abrasive action of wave transported boulders, pebbles and sand causes rapid erosion of the coastal rock. The nature and structure of the rocks forming a coast are, not infrequently, the most important factors which determine the detailed morphology of the coast. For this reason, geological factors may rank equally with marine factors in coastal development.

The simplest type of coastal landform is the cliff, which may have a wave cut platform at its base (Fig. 5.17), or the platform may be absent (Fig. 5.18). Cliffs develop by the cutting of a notch at the base by wave action, which leads to

Fig. 5.17 Cliff with wave cut platform, south coast, New South Wales, Australia.

Fig. 5.18 Cliff without platform, south coast, New South Wales, Australia.

collapse of the overlying rock, often, but not always, along a structural plane such as a joint. The wave-cut platform lies between high and low tide levels, and is the result partly of weathering, partly of marine erosion.

The width of platforms tends to be self limiting since, as the width increases, the waves traversing the platform lose their energy, and hence their ability to erode. As wave action at the cliff base decreases, the cliffs become degraded by weathering and gullying.

Debris produced by wave action, as well as that brought in by rivers, accumulates near the coast. The debris is moved along the coast by longshore drifting; currents, in themselves, are capable of moving only the finer material. It is the longshore drift, which results from wave refraction that moves the coarser material. An understanding of this movement of sediment by longshore drift is vital in the design of any harbour or beach engineering works. Much of the debris carried by longshore drift is thrown up by waves, and accumulates to form beaches, which are, for the most part, best developed in embayments in the coastline (Fig. 5.19). Material eroded from headlands is carried into bays where it

Fig. 5.19 Aerial view of beach on the east coast of Australia.

constitutes potential nourishment for beaches. During calm conditions beaches undergo construction; under storm conditions, beaches may be modified or even destroyed. Where several periods of beach construction have taken place, each period is represented by a berm.

Those coasts which have gentle offshore gradients may have offshore bars which have been built up by waves breaking some distance from the shore. These gradually move shorewards to become barrier bars or beaches. The barrier bar may form the base for dune accumulation while, behind the bars and dunes, lagoons may be developed. As a result of evaporation these lagoons become salt marshes, and ultimately dry up. Foredunes may be constructed behind a beach due to the action of the wind, which transports sand from the beach, and depositing it where vegetation forms a trap. Several lines of such dunes may be present, parallel to the coast.

5.8 Unstable landforms

The term 'unstable landform' is one which is difficult to define since, in one sense, all landforms are unstable. For the present purposes, the term is applied to those landforms which are unstable with respect to engineering works, i.e. landforms on which some failure will certainly occur within the lifetime of an engineering structure. Such landforms include active volcanoes, hillslopes which are actively moving, and areas of reactive soils. It would be possible to include such features as areas of fill and areas subject to subsidence as a result of mining operations, but these are excluded here on the basis of their not constituting natural geomorphic features.

The effects of volcanic activity are well known. The landscape can be changed dramatically by eruptions, which are accompanied by earthquakes, landslides and rock avalanches. Ash and lava produced by the activity also modify the landscape. The potentially unstable hillslope is marked by terracettes (Fig. 5.20) and by mudflows (Fig. 5.21) and landslides. The engineering

Fig. 5.20 Terracettes on hillslope, New South Wales, Australia.

Fig. 5.21 Mudflow on hillside, New South Wales, Australia.

significance of such slopes is discussed later; however, it can be said here that any engineering activity on these slopes will almost certainly lead to renewed movement. Areas of so-called reactive soils are occupied by soils which are expansive, and which, frequently, especially in arid and semi-arid areas, are unsaturated. The soils expand on wetting and shrink on drying, setting up stresses in structures founded on them. These stresses can produce damage, or even failure, in rigid structures.

5.9 Geomorphic maps for engineering purposes

The delineation and representation of geomorphic features on a map is now an essential part of an engineering geological investigation. These maps show, in varying detail depending on scale, such features as streams, springs, landslides, slope gradient, and lines of slope change (Fig. 5.22). The geomorphic map may be

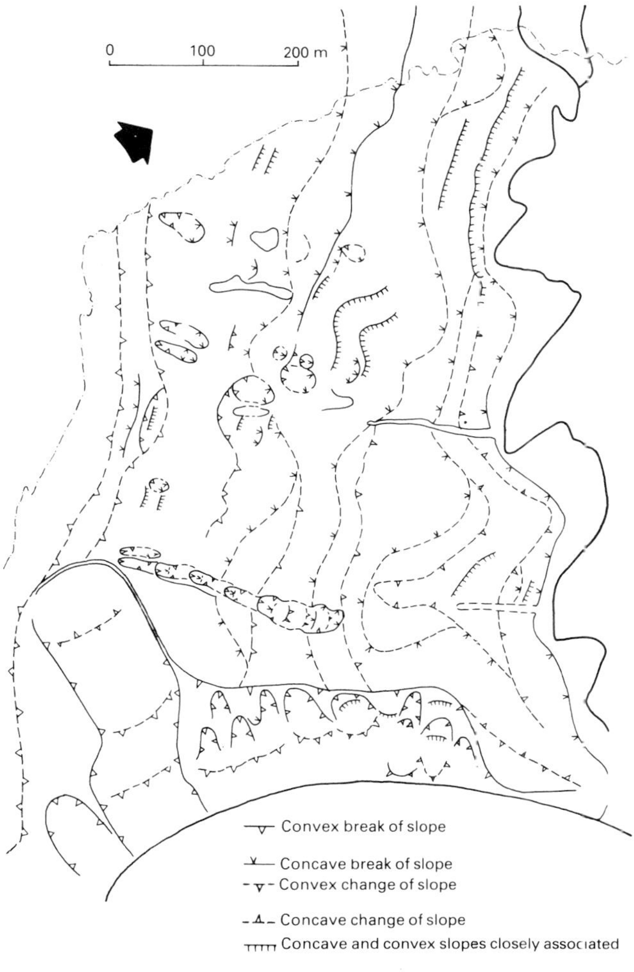

Fig. 5.22 Geomorphic map of landslide area, Lorne, Victoria, Australia. [After Evans (1973).]

partly geological inasmuch as accumulations of lake and other debris are defined, and such features as caves and sinkholes are located. If appropriate, the geomorphic map is accompanied by a slope category map (Fig. 5.23). These define areas of similar slope gradient.

Of course, geomorphic maps are used for areas other than slopes. For example, they may be used to delineate various erosional and depositional

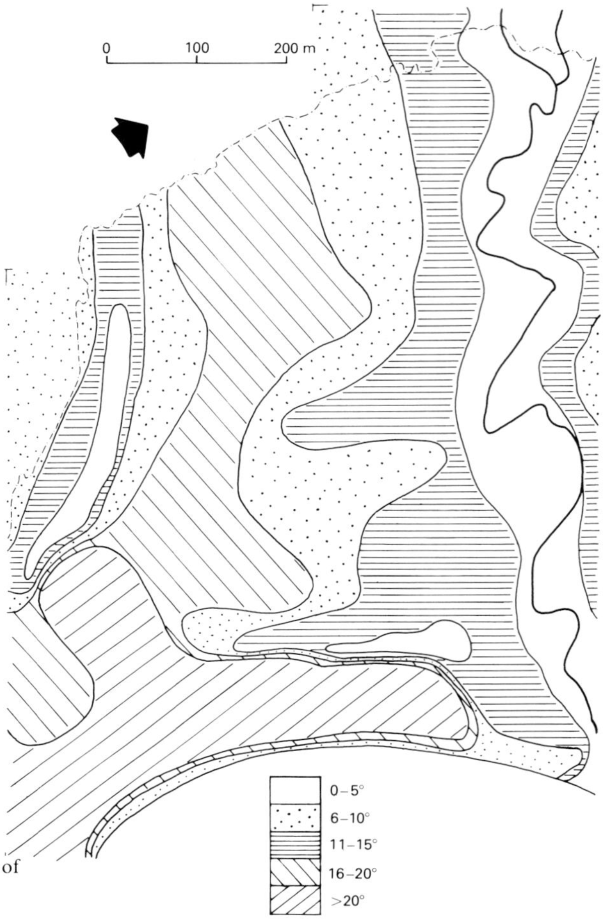

Fig. 5.23 Slope category map of landslide area, Lorne, Victoria, Australia. [After Evans (1973).]

features, and stream bed characteristics along a stream channel, and, in arid regions, areas of stony desert and of dune accumulation.

The Geological Society's report "Land Surface Evaluation for Engineering Practice" (1982) discussed the application of geomorphological principles to engineering geology. Land surface evaluation tends to complement terrain evaluation which is discussed in Chapter 8. Land surface evaluation involves a number of systems and techniques depending on the purpose of the evaluation. These include: land classification, land surface mapping, air photograph interpretation, and the use of remote sensing systems additional to aerial photography. Geomorphological maps produced can be of different types, such as morphological, which present the detailed shape of the land surface, and which are completely factual, without any interpretation. Data on these maps — slope gradient, relief amplitude and stream channel characteristics — can be directly related to engineering aspects of the project. Morphochronological maps distinguish the various landform units according to their age. A distinction is made between recent landforms, and relict forms related to past processes. Much of the information on this type of map may be interpretative.

Morphographical maps depict the landform units by their current shape, and are identified by a specific name, e.g. fan; landslide. Finally, the morphogenetic map depicts the origin and development of the landform units. These maps include the nature of the materials, and data on the processes currently developing the topography. It may, in some cases, be necessary to produce separate maps to emphasize some one aspect of interest to the engineer, e.g. waterways, landslides, or other potential hazards.

Geomorphological maps have application to a number of engineering related undertakings — planning and design of projects, and materials investigation. They have considerable application, especially in developing countries, to regional and town planning.

References and further reading

Chorley R. J. (ed.) (1971) *Introduction to Fluvial Processes.* Methuen, London.

Davies J. L. (1972) *Geographical Variation in Coastal Development.* Oliver & Boyd, Edinburgh.

Davies J. L. (1969) *Landforms of Cold Climates.* Australian National University Press, Canberra.

Evans R. S. (1973) Slope stability studies. Unpublished Honours thesis, University of Melbourne.

Gregory K. J. & Walling D. E. (1973) *Drainage Basin Form and Process.* Edward Arnold, London.

Leopold L. B., Wolman M. G. & Miller J. P. (1964) *Fluvial Processes in Geomorphology.* W. H. Freeman, San Francisco.

Mabbutt J. A. (1977) *Desert Landforms.* Australian National University Press, Canberra.

Quarterly Journal of Engineering Geology, Vol. 12, No. 3. London, 1979. Symposium on Engineering Geological Mapping for Planning, Design and Construction in Civil Engineering.

Twidale C. R. (1968) *Geomorphology.* Nelson, Sydney.

Twidale C. R. (1976) *Analysis of Landforms.* John Wiley, Sydney.

Water Research Foundation of Australia (1976) *Unstable Landforms.* Sydney.

Young A. (1972) *Slopes.* Oliver & Boyd, Edinburgh.

6

Soil: Material and Mass

6.1 General statement

With the exception of purely organic soils such as peat, all soils are derived from rocks; the type of soil is a function of geology, climate and the general environment. The properties of soils are distinctly different from those of rocks: an area exists, however, where it is very difficult to decide whether a material is a rock or a soil. Both rock and soil can be defined as aggregates of mineral particles, with voids which are occupied by either, or both, air and water. Rock material is generally firm, compact, and solid, with a relatively high strength and possessing a certain elasticity. Soils, on the other hand are generally loose aggregates, with low strength, and with little or no elasticity, and the individual particles can be separated quite readily. Moreover, the water in the voids plays a much greater part in determining the properties of a soil than it does those of a rock. The term soil itself means different things to different people. To the pedologist it is a profile comprising three separate horizons, each of which is important in itself and in the total soil. The geotechnical engineer tends to disregard the A horizon of the pedologist, because of its inferior engineering characteristics, and regards as 'engineering soil' the completely weathered grade VI rock. The A horizon, often containing organic matter, should not be ignored because it does have an important role in many engineering situations, especially in slope instability.

The A horizon of the pedologist is a zone of eluviation or leaching, the upper part of which contains much organic material under many conditions. The B horizon is a zone of illuviation or accumulation, while the C horizon is weathered rock. The actual profile developed depends on the parent rock and on climate (Fig. 6.1). Even the so-called transported soils will, with time, develop a profile under the influence of climate.

In the broadest genetic sense, soils can be grouped into two classes for engineering purposes:

(1) Residual or *in situ* soils, which have formed, in place, by the weathering of a rock mass. Such soil masses are, within the limits set by the natural profile, often homogeneous; certainly, the C horizon will be homogeneous if the rock mass itself is homogeneous; these soils reflect the characteristics of the rock mass from which they have been derived.

(2) Transported soils, which have been transported by wind, water, or ice, and deposited to form a soil mass. Some soil masses of this type may be homogeneous, but most are extremely inhomogeneous. Whereas the

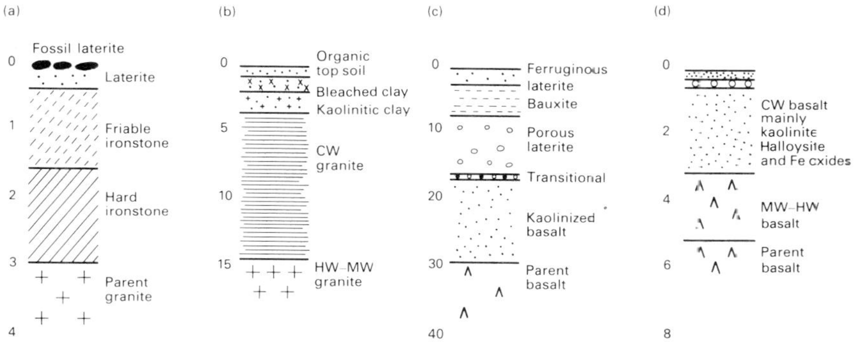

Fig. 6.1 Profiles of soils developed under differing climatic conditions: (a) Tropical lateritic soil. Dohomey, Africa, derived from granite. [After McNeil (1964).] (b) Temperate zone soil. Bulla, Australia, derived from granite. (c) Tropical lateritic soil, on basalt, Southern India. [After McNeil (1964).] (d) Temperate zone soil on basalt, Kiama, Australia.

composition, texture and fabric of residual soils are determined by the parent rock mass, climate and agents of weathering, these characteristics of transported soil are determined, in addition, by the agent of transportation, the distance which they were transported, and the conditions of deposition. Furthermore, they have been derived from more than one particular rock mass.

The mechanical properties of the soil material, and ultimately of the soil mass, are fixed by the mineralogy, particle shape, texture, fabric and moisture. The material properties are of prime interest if the soil is to be used as a construction material; if the soil mass is to form the foundation of a structure, it is the mass properties which demand attention. The properties of soil in the disturbed condition differ strongly from those in the *in situ* condition: this fact emphasises the importance of the soil fabric in determining the soil's properties since, with disturbance, the fabric is totally destroyed, and a new fabric developed.

6.2 Mineral composition of soils

The mineral components of a soil are fixed, in the first instance, by the nature of the parent rock, modified by the degree and nature of the weathering, which are determined by climate. Soils formed from similar rocks, but under varying climatic conditions, may have quite different mineral compositions, especially in so far as the clay minerals are concerned. Under certain climatic conditions, e.g. in arid regions, some other minerals such as gypsum and halite may be present to influence the properties of a soil; however, it is the clay minerals which have the greatest effect on the mechanical behaviour of a soil. Organic matter, usually present in poorly drained soils in wet areas, has a deleterious effect: for this reason, organic rich soils are avoided as both foundations and as construction materials.

6.2.1 CLAY MINERALS

Since it is the structure of a clay mineral which controls such aspects as ion exchange, and the admission and expulsion of water from the crystal, it is the structure which is primarily responsible for the influence which these minerals exert on a soil.

The *kaolinites* form very stable soils since their tight, non-expandable structure resists the introduction of water into the lattice, and hence the soils do not incur the destabilizing effects of water. Furthermore kaolinites, when wet, are only moderately plastic and tend to have a higher coefficient of internal friction, and therefore a higher shear strength, than other clay minerals. Generally, unless impurities are present, kaolinites are not subject to expansion with wetting. One of the kaolinite group, halloysite, occurs as rounded or flattened tubes; if water is introduced, the tube-like structure is capable of creep and flow under the lubricating action of the water. This has been likened to a 'roller bearing' effect.

In the *montmorillonites,* the sheets making up the structures are loosely bonded, so that an unstable mineral results, with the instability increasing if water enters the structure, which it can, quite readily. Water molecules insert themsleves between the loosely bonded sheets, causing expansion of the structure. When this occurs, the montmorillonite sheets are enclosed in films of water, and so the soil has high plasticity and a low coefficient of internal friction. The structure can lose water as readily as it can take it up; with water loss, montmorillonite soils are subject to severe shrinking and cracking.

The *illites* are similar to the montmorillonites, but the crystals tend to aggregate and, as a result, the aggregate structure tends to attract less water. In comparison with the montmorillonites, the illites have a much more limited hydration capacity, the swelling–shrinking characters are less marked, and the coefficient of friction is somewhat higher.

The cation exchange capacity, and the nature of the exchangeable cation are two factors which play a part in determining the mechanical properties of a soil. The commonly exchangeable cations in a clay soil are Na^+, K^+, Ca^{2+}, Mg^{2+}, H^+ and NH^+_4. Clays with H^+, but particularly with Na^+ tend to possess poor mechanical characteristics, and tend to be unstable. A marked improvement can be achieved if the H^+ or Na^+ are exchanged for Ca^{2+}. The nature of the exchangeable cation, and the cation exchange capacity are the bases for soil stabilization using chemical substances, and for electro-osmotic processes. It is known, too, that these aspects of clay minerals can affect the stability of earth dams, especially with respect to piping failure (Wood *et al* 1964; Beavis *et al* 1978). Water in storage may contain ions in solutions which undergo exchange reactions with clay minerals in the soil of the dam. This becomes critical when the water is saline and Na^+ of the water exchanges with other cations on the clays.

Clay minerals may be present in soils in a number of ways, depending to some extent on the origin of the soil. In residual soils, the clay minerals tend to occur as aggregates, pseudomorphous after the mineral from which they formed during weathering. In this way, clay aggregates occur after feldspar, olivine, micas, amphiboles and pyroxenes. On the other hand, in transported soils, clay minerals may occur as much smaller units, often, but not always, uniformly distributed throughout the soil mass. The mode of occurrence of clay minerals in the soil, and their distribution, can affect the mechanical and hydrological properties of the soil.

6.2.2 OTHER MINERAL CONSTITUENTS

Mineral constituents other than the clays may be rock fragments, quartz, feldspars, micas, amphiboles, pyroxenes, olivine, iron oxides and accessory minerals, resistant to weathering, such as garnet, zircon, sphene, apatite, rutile and tourmaline. It has been noted that halite and gypsum are commonly present in arid zone soils. Although these last two minerals can affect soil properties, overall, the clay minerals are the most important. The significance of the main rock-forming minerals lies more in their size, shape, and distribution than with their chemical nature. The influence which quartz, feldspars, and the other rock-forming minerals exert on soil is, as yet, imperfectly understood. It is known that the shear strength of a soil can be modified by the nature of the silicate mineral present: the coefficient of friction of quartz on quartz differs from that of feldspar on feldspar, or of quartz on feldspar. Whether this is important remains uncertain. Soils with a high proportion of highly cleaved minerals may develop an anisotropy not present in soils poor in such minerals. The structure of these minerals, which may determine the strength of bonding with clay minerals, will have some influence, albeit minor, on the soil's characteristics.

6.3 Texture of soils

In engineering, the term *texture* is rarely used; it is more common to refer to the grading of a soil. However, texture implies more than particle size distribution, and gives some consideration to the shape of the grains. Texture, together with fabric, are the most important physical characteristics of a soil in the determination of its engineering properties. Any one soil is usually made up of a number of different sized particles. Textural classes have been defined (Table 6.1)

Table 6.1 Textural classification of soil particles.

Class	Particle size range (mm)
Gravel	>2.00
Coarse sand	2.00–0.20
Fine sand	0.20–0.02
Silt	0.02–0.002
Clay	<0.002

as gravel, sand, silt and clay. A soil classed as a sand, for example, will not necessarily contain exclusively sand-sized particles. It will usually contain silt and clay-sized particles, but in lesser amounts than sand-sized grains. Similarly, a clay may contain a certain number of silt-, and even sand-sized particles. The size distribution aspect of soil texture is usually represented on a grading curve (Fig. 6.2). The curve gives a visual representation of the proportions of each size class present; additionally, the grading of the soil can be assessed. A uniform, or

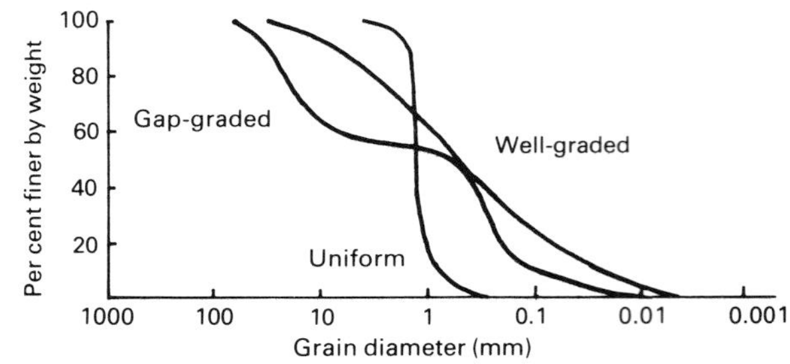

Fig. 6.2 Grading curves for three soils showing characteristic features of uniform or poorly graded; well graded; and gap graded soils.

poorly graded, soil contains a high proportion of one particular size class; a well graded soil contains roughly equal proportions of all sizes. Other important parameters can be determined from the curves, e.g. the *effective size* (d_{10}), the maximum diameter of the smallest 10%; and the *uniformity coefficient*, d_{60}/d_{10}, which is the ratio of the maximum size of the smallest 60% to the effective size. The uniformity coefficient is higher in poorly graded soils than in well graded soils, having a maximum value of one for a perfectly uniform soil; it tends to be higher in sands than in clays. These two parameters, and especially the uniformity coefficient, are most important expressions of the grading of the soil.

The mineral composition of the various size classes varies. Figure 6.3 is based

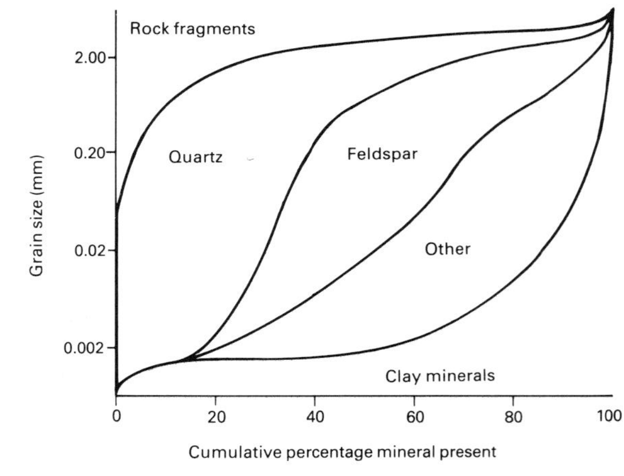

Fig. 6.3 Mineral composition of various grades of a residual soil, developed on granodiorite, Bulla, Victoria, Australia.

on the analysis of a residual soil, developed on a granitic rock, classed as a well graded sand. It can be seen that the clay fraction is not constituted entirely of clay minerals, but contains some quartz, feldspar and micas. With increasing grain size, quartz and rock fragments become dominant. The proportion of rock fragments in a soil, especially one derived from a crystalline rock, is a measure of the degree of weathering of the soil. The more advanced the weathering, the greater the granular disintegration, and hence the proportion of monomineralic grains will increase, and that of rock fragments decrease.

The shape of the grains is to be regarded as a textural feature of the soil. Shape is determined by the processes of soil formation, and, especially in residual soils, by the nature of the parent rock.

For transported soils, shape is determined largely by the nature of the transporting medium, and the distance transported. Shapes of grains can vary from platy to round (Fig. 6.4), and determination of the dominant grain shape is essential for the full description of soil texture. A soil composed of smooth, well rounded grains will respond to stress in a quite different manner to one composed of angular grains, where interlocking is possible. A soil which possesses a high proportion of platy particles may have a strong planar anisotropy which, again, will influence the behaviour of the soil under stress.

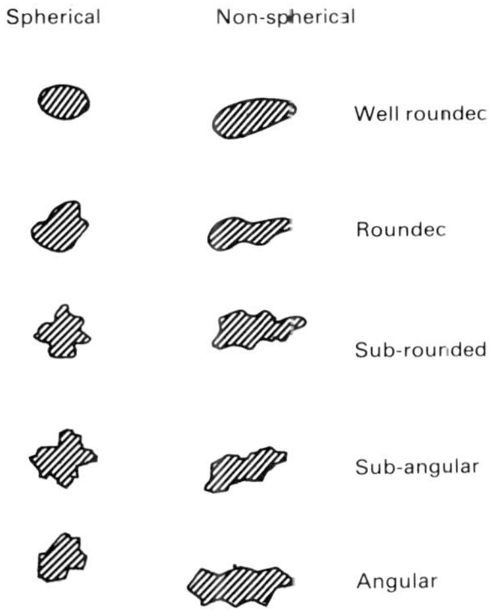

Fig. 6.4 Shape terms for soil particles.

6.4 Soil structure and fabric

By structure of the soil is meant the arrangement of the grains which constitute the material; by fabric is meant the three-dimensional spatial orientation of grains and voids. Three classes of structure may be distinguished (Fig. 6.5):

(1) cohesionless soil structures, e.g. sand;
(2) cohesive soil structures, e.g. clay;
(3) composite soil structures, e.g. sandy clay;

although, as will be shown in Section 6.5, some doubt exists regarding the validity of this simple classification. Dry granular soils, such as a sand, have an unstable structure, which is readily destroyed. If moist, sand grains may hold together due to an apparent cohesion, but on drying out, the structure again becomes unstable. If the grains of a sand are cemented, the structure is stable, but in this case, the material is approaching a sandstone. The structure, and the fabric, of sands in the undisturbed condition, is a function of the dynamic processes involved in deposition.

The structure and fabric of clay soils reflect the origin of the soils. A residual clay soil has much the same structure and fabric as the parent rock. Inorganic clay soils, formed by sedimentation, often have a flocculant structure. If

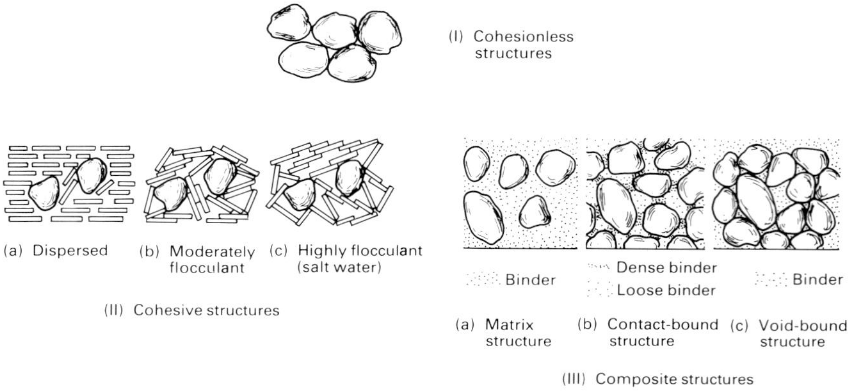

Fig. 6.5 Typical soil structures. Note that some doubt has now been cast on this classical view of soil structure following work on fabric analysis and detailed studies using scanning electron microscopy.

electrolytes are present, these permit flocculation of the particles to produce large aggregates and large voids. Laminated clays are usually the result of precipitation in still water. Osipov and Sokilov (1978) have examined the structure of clay soils using scanning electron microscopy, and have been able to show that each genetic type of clay soil has its characteristic microstructure (Fig. 6.6), although they show that past depositional history can affect the structure (Table 6.2).

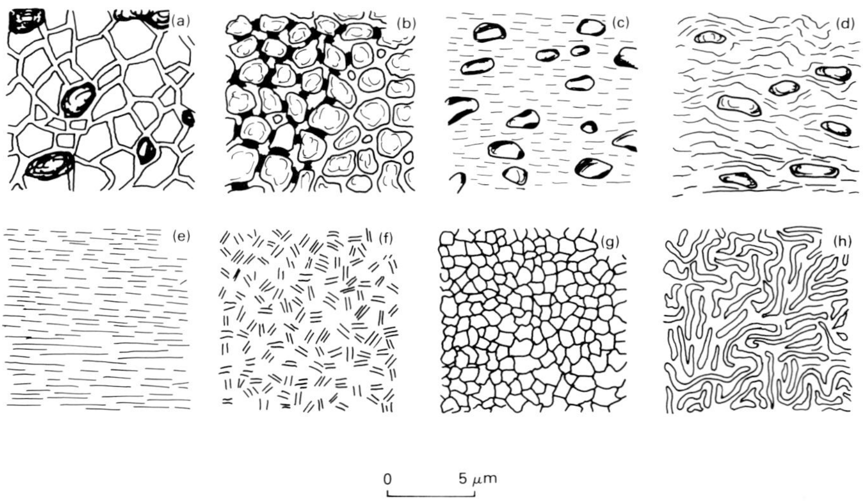

Fig. 6.6 Models of the structures of clay soils. [After Osipov and Sokilov (1978).] (a) Honeycomb. (b) Skeletal. (c) Matrix. (d) Turbulent. (e) Laminar. (f) Domain. (g) Pseudoglobular. (h) Sponge.

Clay soil structures, in many instances, are unstable and can be readily changed. Swelling and shrinking are characteristic manifestations of such changes, but flocculation is as important, if less obvious. The rapid destruction,

Table 6.2 Structures of clay soils.

Genetic class	Structure: Recent	Structure: Compacted	Orientation
Marine	Honeycomb	Matrix and composite. If high degree of compaction, turbulent or rare laminar	Clay particle preferred orientation poor in recent clays but increases with compaction
Estuarine	Matrix	As for marine	As for marine
Lacustrine	Honeycomb	Matrix	Medium to high
Glacial	Skeletal-Matrix	Matrix	None
Glacio-lacustrine	Laminar	Laminar	Very high
Residual	A wide variety of structures, but with pseudoglobular and sponge structure common. Structure depends on parent rock, e.g. kaolinitic clays on granite have domain structure, while nontronite and montmorillonite on basic rocks have globular and sponge structure.		

or alteration, of the structure and fabric of a clay soil by man-made forces is referred to as *remoulding* of the clay. The *sensitivity* of a clay soil to remoulding is measured by the ratio of the uniaxial compressive strength before remoulding to that immediately after remoulding. Sensitivity varies from 1 for heavily overconsolidated clays, to 100 for extrasensitive, or quick, clays. Quick clays may have their structure altered by remoulding or they may lose it completely.

The loss of structure of extrasensitive clays is not to be confused with thixotropy. When a thixotropic clay soil is subject to vibration, it liquifies without the addition of water. In effect, the soil loses all of its shearing resistance. After a period free from vibration, the liquid soil regains its shear strength as it changes back to the solid state. Thixotropy is probably due to a weak structural bonding within the clay which breaks readily under dynamic forces, but which can be re-established quite readily when the application of these forces cease.

Soils other than clays are affected by reworking due to complete change in structure and fabric, and this results in important changes in the mechanical properties. For example, a residual soil, classed as a well graded sand, developed on a granodiorite, gave values for permeability of 2.458×10^{-7} m s^{-1}; cohesion of 18.53 kPa and coefficient of internal friction of 30° on the undisturbed soil. The sample was thoroughly reworked, recompacted to the undisturbed density with the same moisture content as the undisturbed soil. The reworked material had permeability of 1.376×10^{-7} m s^{-1}; cohesion of 27.9 kPa, and coefficient of internal friction of 28°.

When any soil is immersed in water it may exhibit a *slaking* behaviour. This involves a total loss of structure, and with this, complete disintegration. The outer parts of the immersed soil become saturated and prevent the air inside the material from escaping. If, in addition, expandable clay minerals are present, destruction proceeds very rapidly. Kaolinite, halloysite, and the clay micas resist the penetration of water, and remain intact when submerged. Montmorillonites and illites, however, show a high degree of slaking.

6.5 Analysis of soil structure and fabric

The principles of the structural analysis of soils were established by Lafeber (1966) and his associates, who investigated the quantitative aspects of soil structure and fabric, and the importance of these in determining the mechanical behaviour of a soil. The basic principle of these studies is that the individual components of a soil, recognized as skeleton grains, clay platelets, and pores (Fig. 6.7), are spatially arranged in a three-dimensional pattern which exerts a

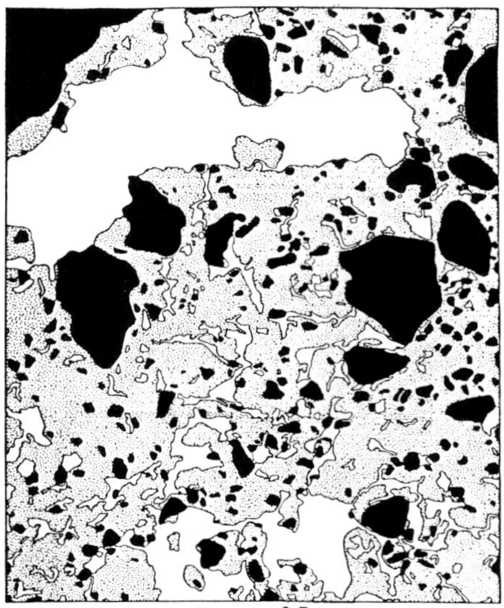

Fig. 6.7 Simplified structural pattern of an arid zone soil, Alice Springs, Australia. The skeleton grains are shown in black; the clay matrix is stippled, and the pores are in white. [After Lafeber (1966).]

definite control over soil behaviour. Structural analysis of a soil is concerned with the description, classification and genetic analysis of the geometrical characteristics of the structure. Analysis considers the distribution of components within a soil, the preferred orientation of the components and a determination of the forces responsible for these.

In a soil, the strengths of these three components are orders of magnitude apart, so that the reaction to stress of each component will be vastly different. The forces acting on a soil are small, so that a preferred crystal lattice orientation of the mineral constituents, due to stress, would not be expected. Reactions to stress are reflected *exclusively* in the preferred *dimensional* orientation of one or more of the components. This is illustrated by Figs 6.8 and 6.9. The first of these figures shows the voids in soil from an unstable slope; the second figure is the equal area projection of the poles to these voids, showing the concentration of poles in relation to the slope. The symmetry is monoclinic–orthorhombic. This can be compared with the projection for a soil from a stable slope (Fig. 6.10),

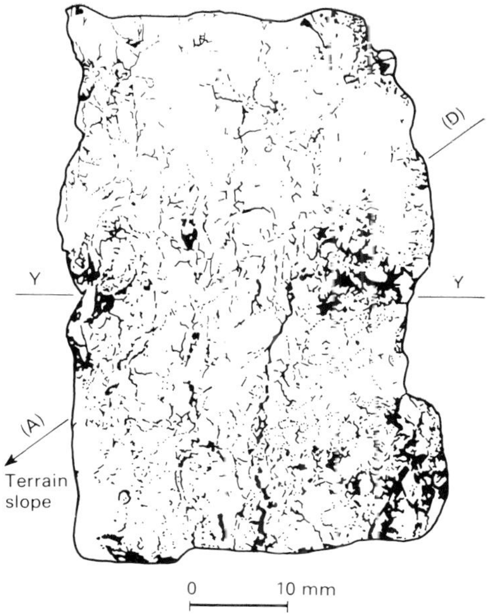

Fig. 6.8 Planar pores in a soil from Olinda, Victoria, on an unstable slope. The soil has developed on Devonian acid volcanic rock. The figure represents a section normal to the strike of the slope. [After Lafeber (1966).]

where the maxima are dispersed, and the symmetry clearly triclinic. Downslope movement has been responsible for the preferred dimensional orientation of the planar voids of the sample of Fig. 6.9. The explanation for the pattern of Fig. 6.10 is more complex, and is, in fact, uncertain.

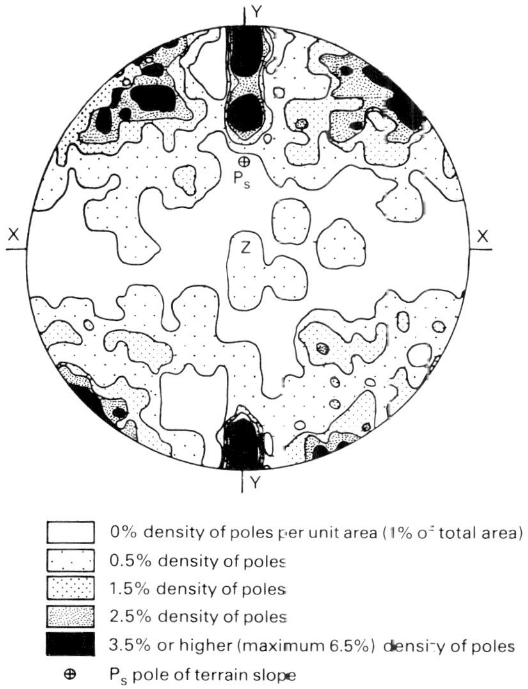

Fig. 6.9 Equal area projection of 196 poles to planar pores in the sample of Fig. 6.8. [After Lafeber (1968).]

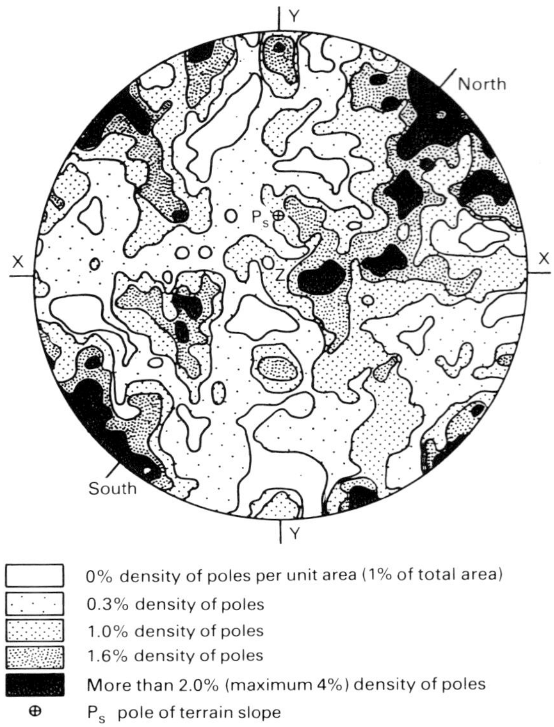

Fig. 6.10 Equal area projection of 318 poles to planar pores in a soil, similar to that of Fig. 6.8, but taken from a stable slope. [After Lafeber (1968).]

One of the most notable results of soil fabric studies has been the casting of doubt on the classical ideas enumerated by Terzaghi. Few, if any, soils appear to approach the condition of a continuous subfabric consisting exclusively of skeleton grains (e.g. flocculant, dispersed); rather, the arrangement is open. Another important result is that most soils are now known to show a preferred dimensional orientation of one, or more, of the components. As a result, the soils have a strong internal geometrical, and therefore mechanical, anisotropy. Such anisotropy will result in a corresponding anisotropy in the internal stress distribution, and therefore in the strain pattern, when the soils are loaded. That is, the strain behaviour of geometrically anisotropic soils will be predetermined largely by the character of these anisotropies.

A further consequence of fabric studies has been the demonstration of the importance of planar voids, or cracks, which constitute discontinuities in the soil mass. Rarely, if ever, is any reference made to these discontinuities in geotechnical studies, and yet they have an importance almost equal to that of fractures in a rock mass.

6.6 Soil consistency

While rocks, irrespective of their moisture content, are always, in practical terms, brittle, soils change their condition with varying moisture content. This is especially true for fine grained soils. Atterberg, a Swedish scientist, developed a method for describing quantitatively the effect of water on the consistency of these soils. Four consistency states were defined: liquid, plastic, semi-solid and solid, and arbitrary but definite limits were set for each state.

The *liquid limit* defines the boundary between the liquid and plastic states. It is the moisture content of the soil at which a trapezoidal groove cut in the soil closes after 25 taps on a rubber plate. The *plastic limit* defines the boundary between the plastic and semi-solid states, and is the moisture content at which a thread of soil 1/8″ (3 mm) in diameter breaks apart when rolled. The *shrinkage limit*, defining the boundary between semi-solid and solid, is the moisture content at which the soil reaches its theoretically minimum value as it dries out from the saturated condition.

In themselves, the Atterberg limits mean very little, but as indices of significant properties of a soil, they are quite useful. The liquid limit has been found to be directly proportional to the compressibility of a soil; while it has been suggested by Sherard (1953) that the relationship between liquid limit and plasticity index can be a guide to the stability of a soil with respect to piping. The *plasticity index* is the difference between the liquid limit and the plastic limit. Other information can be obtained from these simple tests. For example, the first indication that a soil is thixotropic may be obtained when highly anomalous values are obtained for liquid and plastic limits, and especially when, in the test for plastic limit, the soil thread behaves plastically even at near zero moisture content.

The Atterberg limits find their greatest application in the classification of fine grained soils, and in the relationships involving these limits which are useful in correlating soil behaviour with simple test data. For example, the *activity* of a soil is the ratio of the plasticity index to percentage clay size particles, and the *liquidity index* relates the moisture content of a soil to the plastic and liquid limits:

$$\text{Liquidity index} = \frac{\text{Moisture content} - \text{plastic limit}}{\text{Plasticity index}}$$

In effect, Atterberg limits describe the clay–water behaviour of a fine grained soil and describe how their behaviour is affected by non-clay particles.

6.7 Classification of soils

Soils are primarily either residual or transported. Such a simple classification is too broad to serve any useful purpose beyond giving a very indefinite picture of the soil. Numerous classifications have been developed for a variety of purposes, and the most general, cited by pedologists, agriculturalists, and irrigation engineers is based primarily on climate. In this system, soils are classified, in the first instance, on the climate in which they have developed. This 'zonal soil classification' rests on a sound basis, since climate, more than any other factor, has the greatest influence on soil formation. However, this zonal classification gives no information regarding the engineering characteristics of the soil.

A sound engineering classification should take into account texture, structure, fabric and composition. No classification has yet been devised which does take all of these criteria into consideration. The so-called 'unified soil classification' (Table 6.3, Fig. 6.11) is the one most generally accepted by soils engineers. This

Table 6.3 Unified soil classification. (After ASTM D2487-66T.)

Major division		Group symbol	Laboratory classification criteria: Finer than 200 sieve (%)	Laboratory classification criteria: Supplementary requirements	Soil description
Coarse-grained (over 50% by weight coarser than No. 200 sieve)	Gravelly soils (over half of coarse fraction larger than No. 4)	GW	0–5*	D_{60}/D_{10} greater than 4, $D_{20}^2/(D_{60} \times D_{10})$ between 1 and 3	Well-graded gravels, sandy gravels
		GP	0–5*	Not meeting above gradation for GW	Gap-graded or uniform gravels, sandy gravels
		GM	12 or more*	PI less than 4 or below A-line	Silty gravels, silty sandy gravels,
		GC	12 or more*	PI over 7 and above A-line	Clayey gravels, clayey sandy gravels
	Sandy soils (over half of coarse fraction finer than No. 4)	SW	0–5*	D_{60}/D_{10} greater than 4, $D_{20}^2/(D_{60} \times D_{10})$ between 1 and 3	Well-graded sands, gravelly sands
		SP	0–5*	Not meeting above gradation requirements	Gap-graded or uniform sands, gravelly sands
		SM	12 or more*	PI less than 4 or below A-line	Silty sands, silty gravelly sands
		SC	12 or more*	PI over 7 and above A-line	Clayey sands, clayey gravelly sands
Fine-grained (over 50% by weight finer than No. 200 sieve)	Low compressibility (liquid limit less than 50)	ML	Plasticity chart		Silts, very fine sands, silty or clayey silty sands, micaceous silts
		CL	Plasticity chart		Low plasticity clays, sandy or silty clays
		OL	Plasticity chart, organic odor or color		Organic silts and clays of high plasticity
	High compressibility (liquid limit more than 50)	MH	Plasticity chart		Micaceous silts, diatomaceous silts volcanic ash
		CH	Plasticity chart		Highly plastic clays and sandy clays
		OH	Plasticity chart, organic odor or color		Organic silts and clays of high plasticity
Soils with fibrous organic matter		Pt	Fibrous organic matter; will char, burn, or glow		Peat, sandy peats, and clayey peat

* For soils having 5–12% passing the No. 200 sieve, use a dual symbol such as GW–GC.

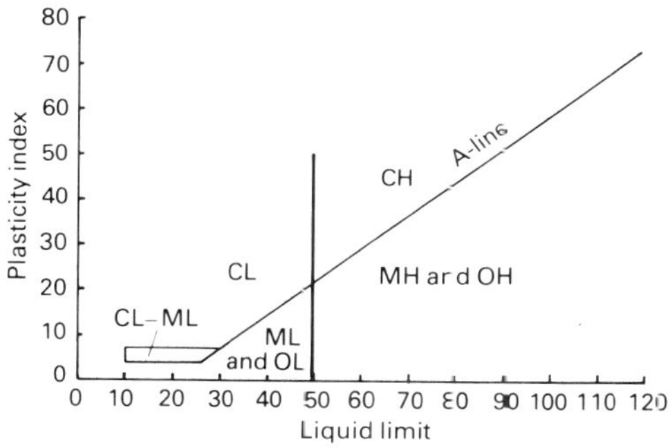

Fig. 6.11 Plasticity chart.

system is based primarily on texture, and takes into account the particle size distribution, effective size and uniformity coefficient. It also takes into account the consistency of the fine grained soils, and of the fine fractions of the coarser grained soils. Such properties as colour, odour and compressibility are also considered for the finer soils.

The classification takes no account of structure, nor of fabric and composition. This deficiency can be rectified by including in the description of the soil a brief statement on composition and fabric, and appending fabric diagrams, just as grading curves are appended to the statement on soil analysis for classification. The very extensive and successful use of this classification suggests that it is almost ideal for engineering purposes.

Another engineering classification, designed by the United States Bureau of Public Roads (Fig. 6.12) is based exclusively on grain size distribution. It is of

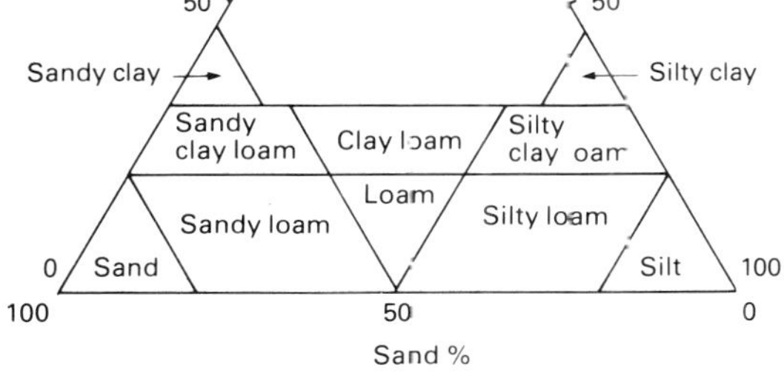

Fig. 6.12 Soil classification scheme of the United States Bureau of Public Roads.

very limited applicability and should be used only for the purpose for which it was designed. Its main advantage is that it permits a brief description of the grading of the soil.

6.8 Mechanical properties of soils

Those properties of soils relevant to engineering geology are the index properties, which are used mainly to classify the soil; the hydraulic properties and the mechanical properties. While the index properties are scalar, the hydraulic and mechanical properties are vectorial, and can vary with direction if the soil material, or the soil mass, possesses a mechanical anisotropy. Unlike rocks, soils possess little, if any, elasticity, and deformation under stress is for the most part non-elastic. Moreover, the response of a soil to stress is influenced to a greater degree by moisture content and air-filled voids than is that of rock.

The index properties of a soil are usually taken to be moisture content, expressed as a percentage of dry weight; the Atterberg limits; porosity (or void ratio); moisture content; and unit weight. The hydraulic and mechanical properties are permeability; unconfined compressive strength; shear strength; compressibility and consolidation characteristics. All of these properties are influenced by geological factors, especially by composition, texture, structure and fabric of both the soil material and mass. In particular, fabric is of major importance since most of the properties are dependent on voids in the soil: the relative amount of void space, the geometrical configuration of voids, their degree of interconnection, and the material filling the voids.

6.9 Characteristics of the soil mass

As was the case with rock, the geomechanical behaviour of a soil mass is a function of its homogeneity, continuity and isotropy. In general, it is true that most soil masses possess these characteristics to a lesser degree than rock masses. Anisotropy in a soil mass is usually less obvious than it is in a rock mass; for this reason it has often been ignored. Nonetheless, moist soil masses do possess some degree of geometric, and therefore of mechanical, anistropy. Residual soils tend to have greater continuity and homogeneity than transported soils which are sometimes notoriously inhomogeneous and discontinuous, especially alluvial soils. In these, both lateral and vertical variations are both frequent and sudden.

6.9.1 HOMOGENEITY

While it is true that residual soil masses *tend* to have greater homogeneity than transported soil masses, the degree of homogeneity will depend on the homogeneity of the parent rock mass. Even so, due to weathering and soil forming processes, the soil mass developed on an homogeneous rock mass will show inhomogeneity which may be gradational. Figure 6.13 shows part of a soil mass developed on an homogeneous quartz biotite sillimanite gneiss in north-east Victoria, Australia, over an exposed section of 10 m. In weathering terms, the exposure is completely weathered rock, the lower section of which shows joints (usually with a kaolinitic infilling), exfoliation surfaces, and well defined foliation. All of these tend to be less apparent in the upper 4 m. Samples were taken at 0.5 m intervals, and the geometric mean diameter of the particles determined; it can be seen that there is a gradual, but definite, decrease in this parameter as the surface is approached. This change is due to increased granular

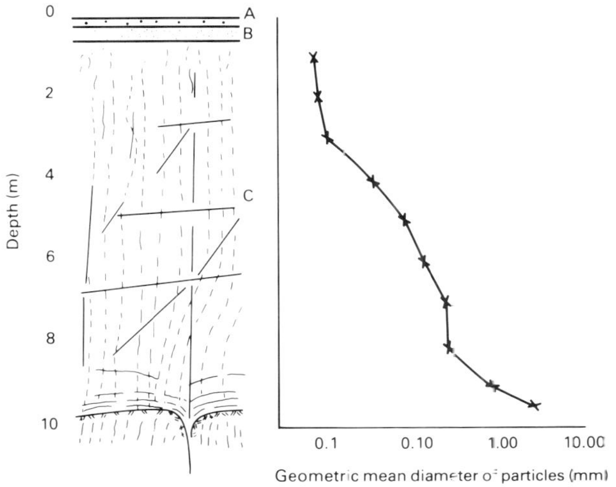

Fig. 6.13 Section in a weathered quartz biotite sillimanite gneiss, Mt Beauty, Australia, showing structures in the soil (C horizon, CW rock) and the variation in geometric mean diameter of particles with depth.

disintegration as weathering becomes more intense. The main point to be made, however, is that this soil mass, derived from an homogeneous rock mass, is strikingly inhomogeneous. Observations suggest that the inhomogeneity observed in this specific example is by no means unusual, and that such behaviour is to be expected. If the rock mass itself is inhomogeneous, greater inhomogeneity will occur in the residual soil, any degree of homogeneity at all is to to be found only in the B horizon of the soil profile.

Two or three examples of transported soils will illustrate the inhomogeneity inherent in these masses. A shallow section of a building excavation at Fisherman's Bend, on the Yarra Yarra River delta, Melbourne, Australia shows the following features. The surface horizon is a coarse sand, 60 cm thick, underlain by 15 cm of highly compressible organic silty clay. Beneath the silty clay is a further bed of sand, below which is an indefinitely thick layer of clays, organic silts and shelly sands. It is known that within a few hundred metres, the thin organic silty clay layer has thickened to over 30 m. Fluviolacustrine deposits which overlie brown coals at Yallourn, Victoria, Australia have a basal gravel with a sandy clay matrix, which is continuous, but of highly variable thickness. This is succeeded upwards by discontinuous lenses of grey clays, brown ligneous clays, mottled, yellow, brown and white sandy clays, sandy silty clays, and silts; the total thickness of the sequence (observed in building foundations) was 12 m. With rare exceptions, no one lens had extent of more than 50 m horizontally.

Alluvial and glacial outwash deposits probably have the greatest inhomogeneity, but flood plain alluvia may show surprisingly high homogeneity, although both lateral and vertical gradations are to be expected. Similarly, wind-blown dune deposits are relatively homogeneous, both vertical and lateral variations being gradational, rather than sharp.

In arid and semi-arid regions, inhomogeneity results often from wind action, although wind-borne deposits themselves may be quite homogeneous. Desert pavements are frequently produced on gravel fans and terraces due to the removal of fine surface material. Beneath the stony surface, fine silt and sand, protected by the pavement of stone, will overlie the gravel deposits which, like all fan and terrace deposits, are excessively inhomogeneous. In some arid zones, transported material, of quite different soil types, overlies the residual soil developed by weathering of the rock mass.

6.9.2 CONTINUITY

In masses of engineering soils, the main forms of discontinuity are:

(1) Boundaries between contrasting textural types. This form of discontinuity is especially important in transported soils.

(2) Inherited discontinuities, such as joints, faults, bedding planes, found only in residual soils.

(3) Planar voids developed in the soil as a result of soil forming processes, and as a result of drying and shrinking. These are normally near-surface features.

(4) Major discontinuities such as tension fissures and sliding planes in soils which have been involved in sliding or slumping.

The proper study of the soil mass requires that these discontinuities be described, with statements on their spacing, extent, orientation and condition included in the description. Such discontinuities affect especially the permeability of the soil mass, but also its reaction to stress. In particular, they can have a strong influence on settlement characteristics and on the ultimate bearing capacity of the soil. Moreover, a particularly well developed set of discontinuities can impart mechanical and hydraulic anisotropy to the soil mass.

6.9.3 ISOTROPY

No soil mass is mechanically isotropic. While it has often been assumed that some transported soils such as beach sands and dune sands were mechanically isotropic, recent fabric studies are indicating at least strong micro-anisotropies resulting from the dynamic processes involved in deposition. Residual soils formed by the weathering of an isotropic rock mass may be essentially isotropic, but, as was shown above, residual soils inherit the anisotropy of the parent rock, except where weathering has been extreme. In some crystalline rocks, exfoliation is an essential phase of the weathering process, and the soils formed carry the traces of exfoliation surfaces. It has been noted, too, that during exfoliation in granites, biotites tend to be oriented parallel to the exfoliation surfaces (Beavis 1962), emphasizing the anisotropy due to the latter.

Anisotropic features of a soil mass have not been studied to any degree, and no certainty exists as to their effect on the mechanical behaviour of the soil mass. It is known, however, that permeability is strongly affected by anisotropy, with considerably higher flow rates occurring parallel to physical anisotropies than in any other direction. Evidence is accumulating that the anisotropy present in clay soils, due to the preferred orientation of clay particles, does have some considerable effect on the mechanical behaviour of these soils, and much research is being directed towards developing a closer understanding of the effects of this anisotropy.

6.10 Geomorphology and soils

The type of soil developed in an area is a function, primarily, of geology and climate. However, the soil is also strongly influenced by the geomorphological regime. On a regional scale it is frequently to be observed that boundaries of major soil units and those of geomorphic provinces coincide. The relationship is, in a sense, to be expected since both soils, and geomorphology, are determined by the same two factors. On a smaller scale, the relationship between soil and geomorphology is sometimes obvious; for example, alluvial transported soils invariably occur on river flats and terraces, and aeolian sands almost invariably constitute dunes. In many other instances, the relationship is not so apparent. In particular, the geometry of uplands, and the form and steepness of slope are highly significant.

Soil variation on uplands and slopes is due, to a considerable extent, to hydrological factors. On flat uplands, run-off is low, or even zero, and infiltration is high. Because of this, rock weathering is highly active, with the removal, in solution, of soluble salts produced by weathering processes; in addition, erosion is minimal. Such is not the case for sharply ridged uplands. On these, and on slopes, run-off tends to be high and infiltration low. Rock weathering is impeded and erosion accelerated. On slopes, too, a downward movement of soils occurs continuously, under the action of gravity, resulting in an accumulation of colluvial soil at the base of the slope. If the geology of the slope is non-homogeneous, different soils will develop on each rock type, but with downward movement, these soils become mixed. Soils low down on slopes tend to retain moisture longer, and to have a higher moisture content than those further upslope. This results in a varying composition, especially in the case of iron oxides, which become hydrated.

The steeper the slope, the greater the variation in soil types and soil profiles. On steep slopes, erosion will be more active, and the profiles shallower. Changes in soil morphology on slopes are better understood if the sequence of soils is regarded, in its totality, as a single unit, each member of the sequence being genetically related to those members on either side of it. Milne (1936) introduced the concept of the *catena* for the sequence of soils on a slope. The morphological features of a soil are analogous to the links of a chain, each link depending on those on either side of it. If changes occur in one link, these are transmitted to all of the other links, those immediately adjacent to the change being influenced most noticeably.

Milne illustrated his concept by a sequence along a slope developed on granite. Here the catena sequence, beginning at the top of the slope, and ending on a plain, was skeletal grey loam → deep colluvial red earth → washed sand → silty clay → clayey sand → clay. It is important to note that catena changes are gradational, and that they may be made much more complex by the presence of soils developed earlier under a different climatic regime.

The results of a detailed study of soil–geological–geomorphical relationships for an area of 40 000 ha, in the arid zone of Australia, is summarized on Table 6.4. In this area, both geology and geomorphology have a high degree of variation, but climate is constant. For a given slope with constant

Table 6.4 Association of soils, rock type and landform Fowlers Gap, Broken Hill, NSW, Australia.

Soil type	Geological–geomorphological characteristics	Vertical lithologic composition
Shallow reddish crestal loams	Occur within extensive outcrops of sandstone, quartzite, and dolomite. Restricted to the crests of ridges, ranges and hills.	Thick mantle of angular rock fragments consisting of sandstones, quartzite, dolomite, silcrete, ferricrete which overlie reddish loam to sandy clay loam. Depth of profile is between 10 and 30 cm
Reddish brown hardpan crestal loams	Occur on broad, flat-topped sandstone ranges Not found on sharp ridges.	Top rock fragment mantle is underlain by reddish sandy clay loam (40–60 cm thick), which passes (sharp boundary) through a carbonate impregnated horizon into a mottled, reddish, sandy subsoil, with abundant *in situ* rock fragments. Maximum depth of profile is 150 cm.
Reddish brown plastic clay/clay loam	Occur on the lower colluvial section of the major slopes underlain by quartzites, sandstone and/or shales.	Thick scree of angular quartzite fragments overlies a reddish brown highly plastic colluvial clay loam to clay (30–70 cm thick). Lower part of profile is usually gypsiferous, sandy, extremely stony, and passes abruptly into a highly shattered quartzite and/or shale bedrock. Maximum depth of profile is 230 cm.
Reddish brown calcareous clay loam	Occurs on the middle and lower parts of slopes underlain by dolomite. Sometimes on minor quartzite slopes associated with dolomite.	Surface scree consisting of dolomite, quartzite, and ferricrete fragments, overlies a clay loam/clay calcareous subsoil, which passes into a highly rock-fragment sandy soil. Depth of profile varies from 20 cm to 300 cm or more.
Dark greyish brown calcareous silty loam	Covers the gently undulating plains or stony downs mostly underlain by shales.	Ground surface consists of bands of stony areas, alternating with stone-free bands. Reddish grey to dark brown clay loamy surface soils overlie a greyish, calcareous silty loam. Subsoil grades gradually into weathered shale. Depth of profile is 250 cm or more.
Partly calcareous reddish brown loamy soils	Occur on undulating lowlands with low quartzite/sandstone ridges, mostly underlain by shales, sandstones and quartzites.	Surface mainly covered by stony pavement, which overlies a reddish brown clay loam (10–60 cm thick), passing either gradually into a rock fragment subsoil or abruptly into a shattered phyllitic shale or sandstone–quartzite bedrock.

Reddish brown terrace clay loam	Occurs on flat terraces. Underlain by silty shales and sandstone/quartzite.	Thick surface pavement of subangular to subrounded fragments of quartzite, sandstone, dolomite, silcrete and ferricrete, above a loose reddish brown clay loam (30–60 cm thick), which passes either into a rock fragment subsoil or weathered bedrock. maximum depth of profile is 250 cm.
Reddish brown sandy clay loam developed on unconsolidated sediments	Occurs only on the alluvial plain. No hard rocks occur in the area.	Stony pavement occurs in patches on the surface. Upper part of profile is sandy clay or sandy clay loam, and grades gradually into a sandy clay loam or loamy subsoil. Soil profiles are very deep (>400 cm).
Sandy loam flood plain alluvium; channel gravels and sands	Flood plains and alluvial flats on major creeks. Most extensive in areas underlain by soft phyllitic shales.	Thick sequence of sandy loam, with irregular horizons (10–30 cm thick) of gravelly sandy loam. Alluvium overlies weathered shale or quartzite at a depth of 150–300 cm or more. Loose and random sequence of channel sands and gravels also occur.

geology, the catena was found to be shallow reddish crestal loams → reddish brown clay loam → calcareous red brown loam → red brown sandy clay loam.

The close relationship between soil and geomorphology is one of the basic principles of the techniques of terrain evaluation (Chapter 8). Appreciation of the association is often important for the engineering geologist, especially in preliminary investigations for construction materials, transport route alignment studies, slope stability, soil erosion and hydrogeological investigations.

References and further reading

Attewell, P. B. & Farmer I. W. (1978) *The Principles of Enginering Geology*. Chapman & Hall, London.

Beavis F. C. (1962) The geology of the Kiewa area. *Proc. Roy. Soc. Vict.* **75**, 344–410.

Beavis F. C., Beavis J. C. & Reade L. M. (1978) Engineering geology of small water storage structures in Australian arid regions. *Quart. Jl. Engng. Geol.* **11**, 279–90.

Capper P. L. & Cassie W. F. (1976) *The Mechanics of Engineering Soils*. John Wiley, New York.

Craig R. F. (1978) *Soil Mechanics*. Van Nostrand, New York.

Hunt C. B. (1972) *The Geology of Soils*. W. H. Freeman, San Francisco.

Lafeber D. (1966) Soil structural concepts. *Engng. Geol.* **1**, 261–90.

Marshall, T. J. (1959) Relations between water and soil. Commonwealth Agricultural Bureaux.

McNeil M. (1964) Lateritic soils. *Scient. Amer,* **211**, 96–102.

Milne G. (1936) Normal erosion as a factor in soil profile development. *Nature* **138,** 548–9.

Osipov V.I. & Sokilov V.N. (1978) Structure formation in clays. *Bull. Int. Assoc. Engng. Geol.* **18,** 83-90.

Sherard J. L. (1953) The influence of soil properties and construction methods on the performance of homogeneous earth dams. *U.S. Bureau of Reclamation,* Tech. Memo No. 645.

Sowers G. B. & Sowers G. F. (1970) *Introductory Soil Mechanics and Foundations*. MacMillan, London.

Terzaghi K. (1943) *Theoretical Soil Mechanics*. John Wiley, New York.

Terzaghi K. & Peck R. B. (1948) *Soil Mechanics in Engineering Practice*. John Wiley, New York.

Wood C. C., Aitchison G. D. & Ingles O. G. (1964) Physicochemical and engineering aspects of piping failures in small earth dams. In Water Resources Foundation of Australia and CSIRO: Colloquium on the failure of small dams, Paper 29.

Wu T. H. (1970) *Soil Mechanics*. Allyn & Bacon, Boston.

Yong R. N. & Warkentin B. P. (1966) *Introduction to Soil Behaviour*. MacMillan, New York.

7

Water in Rock and Soil Masses

7.1 Introduction

The presence of water in a static condition in rocks and soils has a considerable effect on their reaction to stress. In any rock or soil mass, however, the water is in a dynamic condition: it is moving. The study of the moving water constitutes the science of hydrogeology. The two most fundamental concepts of hydrogeology are those of the hydrological cycle and of the water table. Hydrogeology is concerned primarily with groundwater as a resource: its availability, chemical and physical properties, and its utility for various purposes such as farm or urban water supply, or for industrial use. Additionally, however, groundwater has to be studied in terms of civil engineering works, e.g. the leakage of water from reservoirs to join the groundwater body; the inflow of groundwater to tunnels; inflow into excavations, and seepage on slopes.

7.2 The hydrological cycle

This concept shown on Fig. 7.1, can be expressed algebraically by the hydrological equation:
where: P is the precipitation; R is the run-off; I is the infiltration; E is the evaporation; and T is the transpiration.

The equation simply states that of the water which fall on the earth's surface, part flows off the surface into rivers, and ultimately to lakes or the sea (run-off); part penetrates into the soil and rock (infiltration); and part returns to the atmosphere by direct evaporation, and by transpiration from vegetation. Infiltration is responsible for groundwater accumulation; evaporation, transpiration and (in part) run-off, for natural groundwater losses

The relative values of run-off and infiltration (and, to some degree, evaporation) are dependent on the geological and vegetative conditions of the catchment on which the water falls. In a catchment which has little soil cover, run-off will be very high; on the other hand, in a well-vegetated catchment, with thick soil and deeply weathered rock, infiltration will be high, relative to run-off. Some water which initially infiltrates into the soil may be lost by evaporation and transpiration, and possibly by groundwater run-off, but much of it will move downwards under the action of gravity, to join the main body of groundwater.

When precipitation occurs, the infiltration rate is highest in the initial stages. However, this rate often decreases exponentially with time, the actual rate depending on the nature and condition of the surface material. Soils which are clayey, and which contain swelling clay minerals rapidly lose their capacity to transmit water due to the closing of cracks and the swelling of clays. On the other hand, materials such as sands do not react as rapidly, but, when saturation is reached, the infiltration rate decreases. In the initial stages of infiltration, the attraction of the water to solid particles by capillary forces is important; when the infiltration has proceeded for a depth of about 1 m, however, the effect of capillary forces in coarse to medium grained soils is quite minor. Capillary attraction is greatest in low moisture content, fine grained soils.

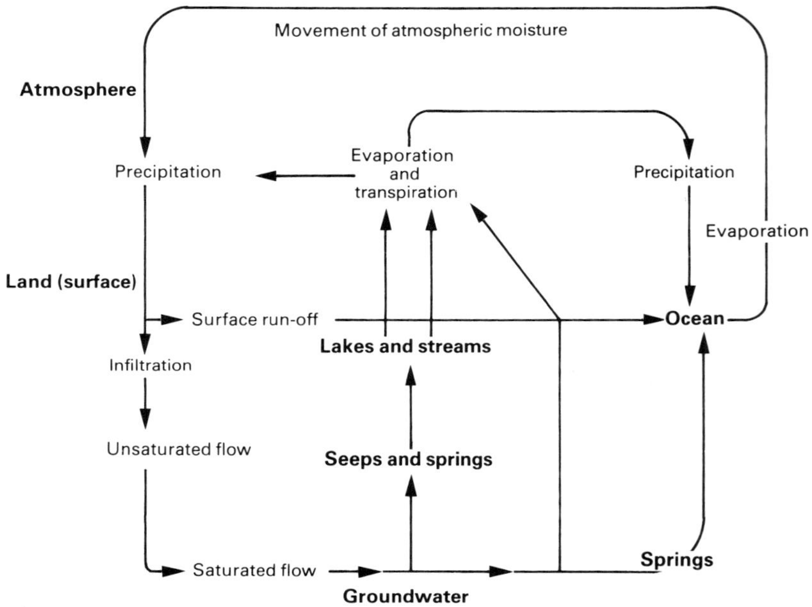

Fig. 7.1 The hydrological cycle.

The total amount of infiltration which reaches the groundwater body is equal to the total infiltration minus the amount of water absorbed by the rock and soil above the water table and that lost by the interflow component of run-off.

The run-off element of the hydrological cycle consists of water which runs directly off the surface, together with interflow, which is infiltrated water moving laterally at shallow depth. Run-off is measured as stream flow, so that total run-off may include some groundwater flow as well as that part of the precipitation which falls directly into the stream.

Under some conditions, part of the run-off may be lost from the stream to the groundwater body. In such a case the stream is said to be influent, as opposed to the effluent stream, the flow of which gains from the groundwater.

Run-off is very strongly influenced by the geological and vegetational conditions of the catchment, both in terms of rate and amount. The amount and rate of run-off for a bare rocky catchment are both very high. For a well vegetated catchment which has thick soil cover and deep weathering, the amount of direct run-off, as well as the rate, are reduced; interflow tends to increase. Bare, rocky catchments tend to produce flash floods after a rainfall event, whereas the soil covered and vegetated catchment results in a more uniform stream flow.

Evaporation and transpiration are usually considered together as evapotranspiration. These processes result in water being returned to the atmosphere in the form of vapour. Unless the water table is close to the surface, groundwater is not directly lost by evapotranspiration. Solar and wind energy are largely responsible for the amount of evaporation; the amount of transpiration will depend on the type of vegetation.

7.3 The water table

Soil and rock contain some water in pores and fractures; however, two general zones may be recognized: an upper unsaturated zone, and a lower saturated zone. The boundary between the two zones constitutes the water table. The upper, unsaturated, rock and soil constitutes the zone of aeration; the lower, saturated zone, the zone of saturation. The water in the zone of aeration can be considered under several categories (Fig. 7.2), although there are no sharp boundaries

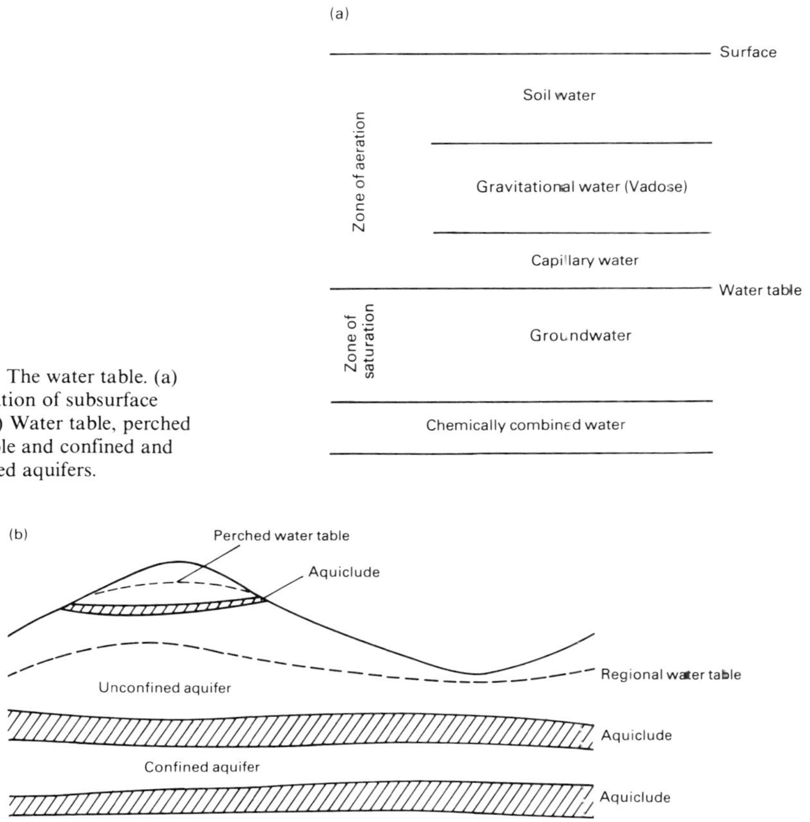

Fig. 7.2 The water table. (a) Classification of subsurface water. (b) Water table, perched water table and confined and unconfined aquifers.

between them. The soil water is subject to considerable variations in both quantity and quality in response to evaporation and transpiration. Vadose water is moving downwards under the action of gravity, and it terminates in the capillary fringe. The upper surface of the fringe is highly aerated, but the lower section is fully saturated.

The water table is a theoretical surface along which the hydrostatic pressure is equal to atmospheric pressure. If a series of wells or bores are drilled to a depth below the water table, groundwater flows into the bores coming to rest at the level of the water table. Water table level is subject to a number of variations due to gains to, or losses from, the groundwater body which are largely seasonal. Minor fluctuations also occur due to tidal influences and changes in atmospheric pressure.

Although, below the water table, all of the rocks are saturated, only a few rocks will transmit water into wells. Those rocks through which water flows relatively readily are termed *aquifers*. Those rocks which do not readily transmit water are termed *aquicludes*. Aquifers may consist of almost any rock type, but either the material, or the mass, must possess a high permeability. If an aquiclude occurs above the water table, *perched* groundwater conditions may develop, with a *perched water table*.

Groundwater which is in direct contact virtually with the atmosphere through the open pores and fissures in the rock and soil is said to be unconfined. Confined water is separated from the atmosphere by impermeable rock and soil. Confined water is sometimes referred to as *artesian* water, though this term should be restricted to water confined in aquifers at sufficient pressure to flow at surface from wells drilled into the aquifer. The pressure of the water in a confined aquifer can be delineated as a surface, analogous to the water table. This is known as the *piezometric surface*.

It is possible, by plotting the levels of water in bores, to construct a contour map of the water table (Fig. 7.3). Similarly, by plotting water pressures in artesian wells, it is possible to prepare a contour map of the piezometric surface. Such maps are of inestimable value in monitoring the effects of groundwater exploitation; they are also of assistance in the planning and design of many civil engineering works.

7.4 Permeability

7.4.1 COEFFICIENT OF PERMEABILITY

Soils and rocks are said to be permeable if the fabric is such that pores are interconnected so that a fluid may flow through the material. It has been established that the flow of fluids through all rocks and soils follows the same laws, so that the difference between the flow of water through a sand, clay, sandstone, or granite is merely one of degree.

As water percolates through a rock or soil, the individual water particles move along paths which follow smooth curves called flow lines. If adjacent flow lines are straight and parallel, the flow is said to be linear. It is assumed, in the

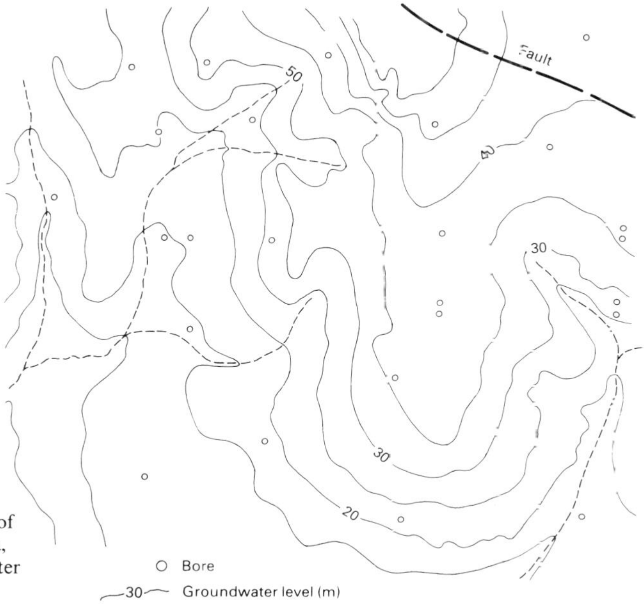

Fig. 7.3 Water table map of area near Darwin, Australia, showing contours of the water table in dry season.

study of permeability, that flow is of this type. Under these conditions, Darcy's Law

$$v = ki \qquad (7.1)$$

(where v is the velocity of flow; i is the hydraulic gradient; and k is a constant, known as the coefficient of permeability) is valid, and applicable. The coefficient of permeability, k, is defined as the rate of flow per unit area of soil or rock, under unit hydraulic gradient; it has the units of velocity. The porosity, shape, size and degree of interconnection of the pores, and the density and viscosity of the water exert the greatest control on permeability. In these terms,

$$k = \frac{V\eta}{i\gamma_w} \qquad (7.2)$$

where η is the viscosity of the water; and γ_w is the density of water.

Permeability has a wide range of values: coarse gravels may have a value of k of 10^5 cm s^{-1}, while dense, crystalline rock material may have k=10^{-12} cm s^{-1}. Because some rock material is virtually impermeable, and flow through the rock mass is usually not so much through the material as along fractures and fissures, it is customary to refer to the permeability of the rock material as the primary permeability, and to that of the rock mass as secondary permeability. Table 7.1 gives some values of primary and secondary permeability.

It can be seen that the primary permeability is up to 10^7 times lower than the secondary permeability. This stresses the danger of applying laboratory values of k, determined on intact rock material, to field problems.

It is now common in practical engineering problems to express permeability of the rock mass in lugeons. This is determined by pump-in tests: 1 lugeon represents a flow of 1 l of water per minute through a borehole of 1 m length

Table 7.1 Coefficient of permeability of rock.

	Primary permeability ($cm\ s^{-1}$)	Secondary permeability ($cm\ s^{-1}$)
Granite	2.3×10^{-10}	1.2×10^{-3}
Sandstone	1.8×10^{-7}	1.4×10^{-2}
Shale	6.8×10^{-7}	3.4×10^{-4}
Phyllite	8.2×10^{-10}	5.6×10^{-5}
Limestone	7.3×10^{-10}	2.4×10^{-3}
Dolomite	4.8×10^{-9}	8.8×10^{-4}

under a pressure of 1 (actually 0.98) MPa. Although a valuable method, experience has shown that this test does not always give reliable or representative results.

Although the differences are not as dramatic for soils, nonetheless the permeability of the soil material may differ by several orders of magnitude from that of the soil mass. Field tests of soil mass permeability are essential for safe and reliable determinations.

7.4.2 FIELD DETERMINATION OF PERMEABILITY

When an unconfined aquifer is pumped, groundwater flows into the well, and a hydraulic gradient is established. Around the pumped well, a zone of rock is depleted of groundwater so that a cone of depletion is developed (Fig. 7.4). If steady state conditions are established, it is possible to utilize data obtained from the pumped well to determine the permeability or hydraulic conductivity of the rock mass. Accurate determination is impossible because of the necessary simplifying assumptions which have to be made, but the values obtained represent a reasonable approximation.

The method depends on measuring the quantity of water pumped from the well, and the determination, by the use of observation wells, of the geometry of the cone of depletion. The necessary assumptions are: (i) the groundwater system is such that it is steady state; (ii) the water table represents the hydraulic gradient; (iii) flow is only towards the pumped well, and in no other direction; and (iv) there are no losses other than from the pumping, and no water gains. It can be shown from Darcy's Law that:

$$k = \frac{Q}{\pi(h_2^2 - h_1^2)} \ln(r_2/r_1) \qquad (7.3)$$

where k is the hydraulic conductivity; Q is the steady state pumping rate; and h_2, h_1, r_2, r_1 are shown on Fig. 7.4a.

In the case of a confined aquifer (Fig. 7.4b), the boundary conditions of the aquifer are determined by the aquicludes; hence, if steady state flow is established in the aquifer, the hydraulic gradient is the slope of the piezometric surface. In this case,

$$k = \frac{Q}{2\pi(h_2 - h_1)t} \ln(r_2/r_1) \qquad (7.4)$$

where t is the thickness of the aquifer.

Other parameters of aquifers, such as transmissivity and storativity, of interest in the development of groundwater as a resource are also determined using pumping tests, with measurements of rate of drawdown with pumping an essential part of the tests.

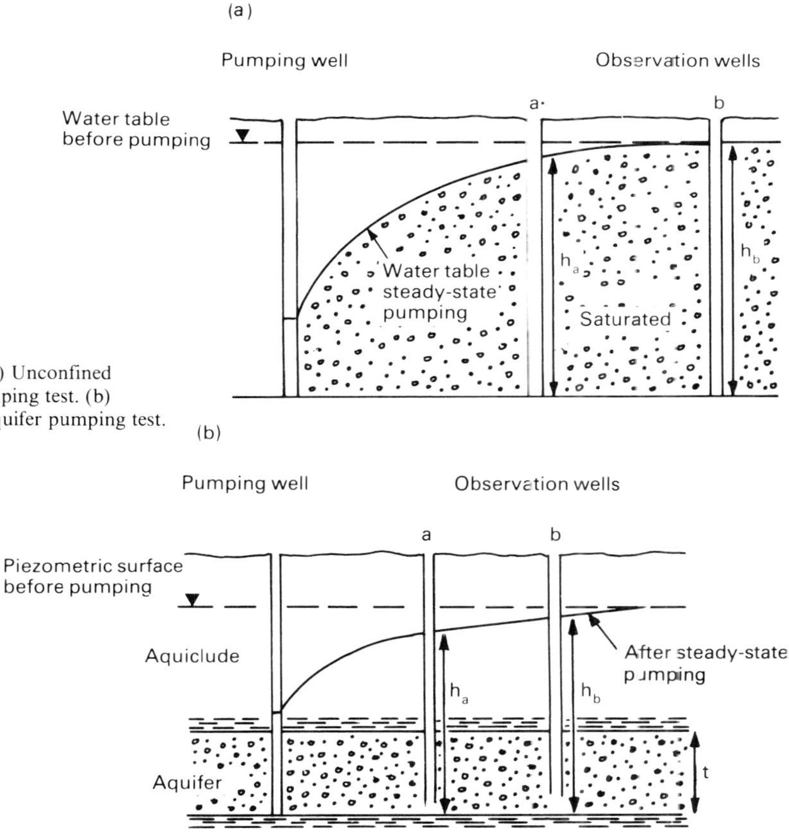

Fig. 7.4 (a) Unconfined aquifer pumping test. (b) Confined aquifer pumping test.

7.4.3 GEOLOGICAL FACTORS AFFECTING PERMEABILITY

Fabric, both on the microscopic and macroscopic scales, is the dominant geological factor in determining the permeability of both soils and rocks. At the simplest level, this can be demonstrated for a soil by comparing the permeability of an undisturbed sample, and a remoulded sample of the same soil Remoulding alters the fabric, and effectively varies the permeability: in most soils, a reduction in permeability occurs with remoulding.

For both rocks and soils, while porosity is essential for flow of water, it is the pore fabric, rather than the absolute volume of pore space which will determine permeability. It is necessary for the pores to be of sufficient size and to be interconnected for flow to occur. Thus a clay, with porosity of 28% had $k = 2.4 \times 10^{-8}$ cm s^{-1}, while a silty sand, with porosity of 16.5%, had $k = 7.6 \times 10^{-4}$ 8 cm s^{-1}.

Fissured clays, with a similar porosity to homogeneous clays, have a coefficient of permeability some 1000 times greater. In rocks, it has been established for sandstones and limestones that a linear relationship exists between porosity and permeability (Fig. 7.5). Rzhevsky and Novik (1971) have

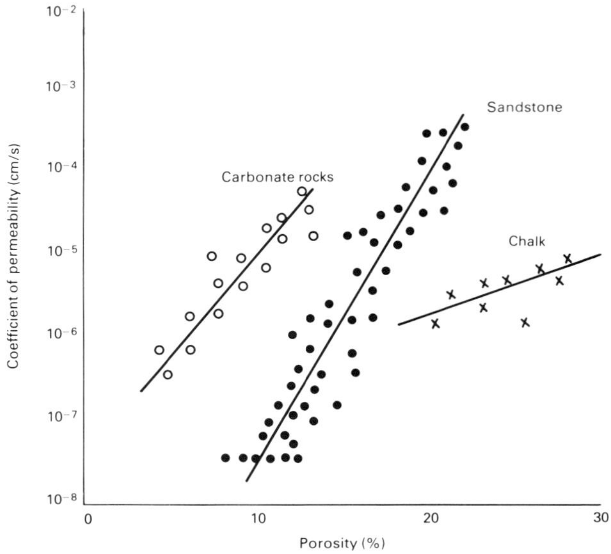

Fig. 7.5 Relationship between permeability and porosity. [After Rzhevsky and Novik (1971).]

shown that for a variety of sandstones, carbonate rocks, and chalky limestone, a linear relationship exists between porosity and permeability. Bell (1978) in a study of sandstones in Northern England found that both porosity and permeability were influenced by the void percentage and grain packing. Particle size distribution did not affect porosity, although it had some influence on permeability. A highly significant relationship was found to exist between porosity and permeability, both of which decrease with increasing depth, although the decrease in permeability with depth does not appear to be as well defined as that in porosity.

It is reasonable to assume that, for rocks which are anisotropic, permeability will be greater parallel to the anisotropy than normal to it. The few results available for both rock and soil tend to substantiate this assumption, but few systematic studies appear to have been carried out.

Pless (1975) systematically examined porosity and permeability of Narrabeen Group sandstones. He showed no significant change in porosity with depth, but a systematic decrease in permeability with increasing depth. Both porosity and permeability were strongly anisotropic (Table 7.2). Permeability parallel to bedding was almost invariably greater than that normal to the

Table 7.2 Porosity and permeability of Narrabeen Group Rocks [After Pless (1975).]

Depth (m)	Rock	Porosity (%)		Permeability ($\times 10^{-6}$ cm s^{-1})	
		Horizontal	Vertical	Horizontal	Vertical
324	shale	4.1	5.0	2.7	2.4
368	shale	3.4	3.1	n.d.	n.d.
393	fine sandstone	8.9	4.6	0.2	0.3
421	fine sandstone	8.0	8.9	0.6	0.2
484	fine sandstone	7.9	8.8	0.2	0.2
508	fine sandstone	9.0	11.0	0.2	0.1
655	med. sandstone	15.5	9.6	0.4	0.6
668	med. sandstone	7.6	n.d.	20.0	9.0
671	coarse sandstone	11.0	n.d.	4.5	14.0
597	coarse sandstone	12.0	13.5	0.1	0.1
606	coarse sandstone	15.0	13.5	0.1	0.1
737	v. coarse sandstone	19.4	14.9	0.1	0.1
674	coarse sandstone	16.0	15.0	39.2	2.2
685	coarse sandstone	12.0	12.0	0.1	0.1

bedding, sometimes by a factor of over 20. Pless's results suggest that no definite relationship necessarily exists between texture, porosity and permeability.

Considerable evidence exists to show that permeability decreases with depth, i.e. with increasing confining pressure. This is due, clearly, to a tightening of fissures, and a closing of microvoids with increasing pressure. A further influence is to be found in decreased weathering with depth. In general, weathering leads to an increase in both porosity and permeability due to the development of microfractures and the opening up of macrofractures.

7.5 Flow nets

The flow of water through soils and rocks can be described by the Laplacian:

$$\frac{\delta^2 h}{\delta x^2} + \frac{\delta^2 h}{\delta z^2} = 0 \qquad (7.5)$$

This equation is represented by two sets of orthogonal lines known as flow lines and equipotential lines. Together, these lines, representing the flow of water through a soil or rock mass, constitute a flow net (Fig. 7.6). The flow lines define the direction of flow; the equipotential lines join points of equal potential head. At the water table, all particles of water have maximum potential. By the time the water has travelled through the rock or soil mass to the equipotential line AB, it has lost potential by the amount of head between the water table and the line A′B′. This head loss has been transferred to the particles in the mass through which flow has taken place. The pressure at any point in the mass due to this transfer is called the seepage pressure. If the pressure is measured at CD, the next equipotential along the flow path, the potential is found to be less than at AB, by an amount h, the head lost. Head is connected to pressure by multiplying h by the density of water. The seepage pressure applied to the particles is then:

$$S = h\gamma_w$$

Although flow nets can be constructed accurately, rapid sketching often

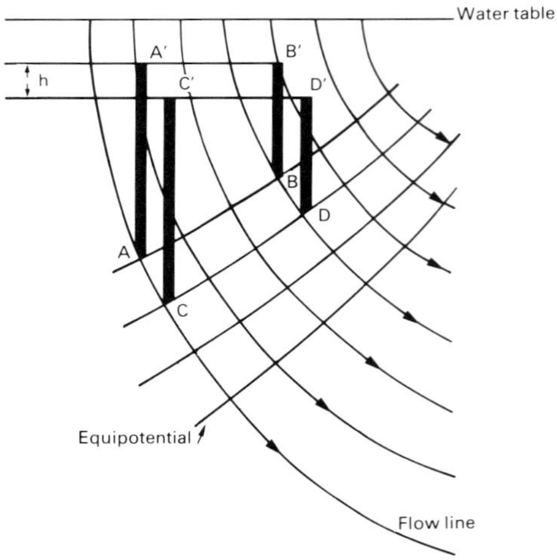

Fig. 7.6 Idealized flow net.

suffices for at least preliminary studies of water movement. The first stage is to consider the boundary conditions: an impermeable boundary constitutes a flow line, while the water table, a water surface, or surfaces of entry and exit of water, constitute an equipotential line. Four boundary conditions exist: (i) the entrance surface, (ii) the exit surface, (iii) the upper flow line, and (iv) the lower flow line.

If the permeability of the soil mass or of the rock mass is known, it is possible, using a flow net, to estimate the loss of water from, say, a storage; it is also possible to calculate seepage pressures. If the flow net has been constructed validly, it can be assumed that, when following any one flow line, the pressure drop between two consecutive equipotential lines is the same. If it is assumed that each square of the flow net represents 1 unit, then the path of water moving from one surface to another consists of n units, and the width of the zone of flow is m units. if H is the total difference in head, the hydraulic gradient is H/n. The flow is then:

$$q = Aki = kH(m/n)$$

The above discussion is based on an homogeneous, isotropic, and continuous mass of soil or rock. Since none of these conditions is normally encountered in practice, the various geological factors will each contribute significantly to affect the flow of water, and hence the form of the flow net. Discontinuities may constitute flow boundaries; equally, they may be the only possible path for the flow of water. The flow net is probably validly constructed only for a uniform mass of soil.

7.6 Water quality

All groundwater contains both anions and cations in solution. The nature and concentration of these ions are of great importance in a number of ways, e.g. the purpose for which groundwater may be used for agricultural or human consumption; and for the effect groundwater may have on concrete footings located below the water table. The nature and concentrations of the ions are largely functions of the geology of the intake areas, and of the composition of the aquifer. Ionic concentrations may be expressed as parts per million (p.p.m.) or, more commonly equivalents per million (e.p.m.). It is usual to plot analyses graphically, so that comparisons can be made readily between different samples. The two commonest graphical methods are the Stiff diagram (Fig. 7.7a) and the Piper diagram (Fig. 7.7b).

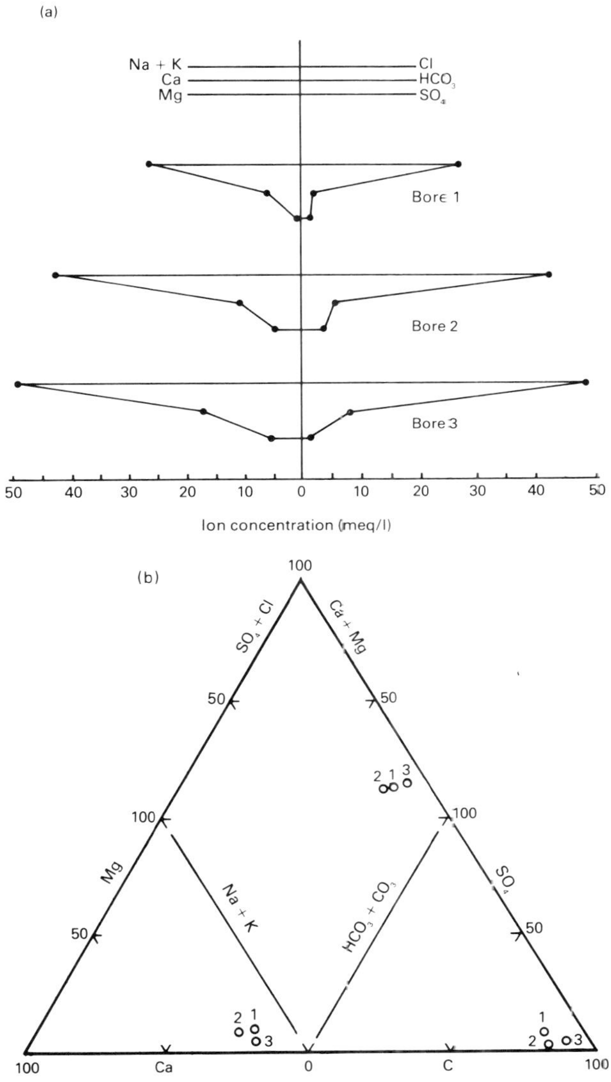

Fig. 7.7 (a) Stiff diagram showing chemical composition of groundwater, Broken Hill, New South Wales, Australia. (b) Piper diagram showing chemical composition of groundwater, Broken Hill, New South Wales Australia.

It is not only chemical quality of water which has to be considered, but also biological quality. Organisms, some of which may be pathogenic, can occur in groundwater which has been polluted by sewage, or decaying animal matter.

Sewage pollution is usually indicated by the presence of *E. coli*. Water which is to be used for human consumption must be free of such pollution.

7.7 Groundwater resources evaluation

The search for groundwater is based on geological data; and of special importance are petrology, stratigraphy, structural geology and geomorphology. Petrology considers the rock types present, and especially their texture and fabric. Stratigraphic techniques are used to define the position and thickness of potential aquifers and aquicludes, and the continuity of these units. Structural geology is utilized to define areas of fracturing which may constitute water bearing units in otherwise impermeable masses, as well as locating aquifer boundaries due to faulting. Geomorphology is indispensable in the study of very young, unconsolidated sediments as potential aquifers, as well as providing valuable data in areas of karst topography.

The basic geological map produced for groundwater studies may be quite different from the conventional geological map (Fig. 7.8) because the interest lies

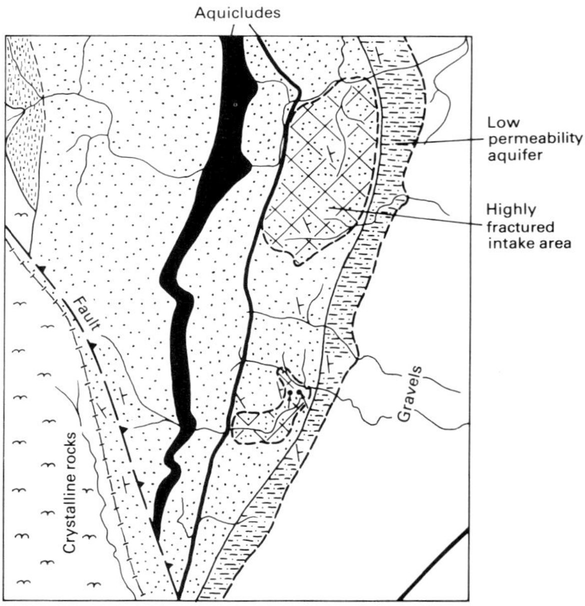

Fig. 7.8 Hydrogeological map, Fowlers Gap, Australia.

in areas of possible groundwater development, aquifers, aquifer boundaries and aquicludes.

The geological studies require as a complement, hydrological studies. Information is required on rainfall, run-off, infiltration and evaporation, so that estimates can be made of the water available for recharge, and of natural discharge into influent streams, and by springs. In conjunction with the hydrological studies, potential or actual recharge areas may be located.

Detailed investigation by geophysical techniques and drilling are used accurately to define aquifers and aquicludes at depth. The boreholes will provide not only geological data, but also hydrological data, permitting the water table level to be determined. Additionally, pump tests will be carried out to determine such important aquifer parameters as hydraulic conductivity, transmissivity, storativity and safe yield. At this stage, samples of water are taken for chemical and biological analysis. Sufficient data may become available to construct, at least preliminary, flow nets of the field. These are not only important in the planning of development, but also serve as a basis for monitoring the field after development.

7.8 Groundwater and civil engineering

To some extent, civil engineers are concerned with the development of groundwater resources, and in particular with that aspect of development involving the hydraulics of aquifers and wells. As this is a highly specialized study, the following discussion will be concerned primarily with groundwater in so far as it is relevant to the construction of civil engineering structures. In this context, groundwater generally ceases to be regarded as a valuable resource, and is considered more as a problem, or nuisance. Any excavation which is carried below the water table in permeable rock or soil, will be subject to inflows of groundwater, and these add to the difficulties of construction. Goundwater also contributes to the instability of slopes, and creates hazards at damsites and in tunnels.

7.8.1 GROUNDWATER IN EXCAVATIONS

All foundation excavations should be sufficiently dry to permit placement of the concrete or other material forming the foundation of the structure. In addition, if an excavation in soil is to be stable, so that the walls stand without support, it is necessary to drain the site. While in small excavations in low to medium permeability materials adequate drainage may be achieved using a sump, for larger excavations below the water table in highly permeable materials, a general lowering of the water table may be required. This is achieved using the *wellpoint* technique (Fig. 7.9). Bores are sunk to a level well below the base of the proposed

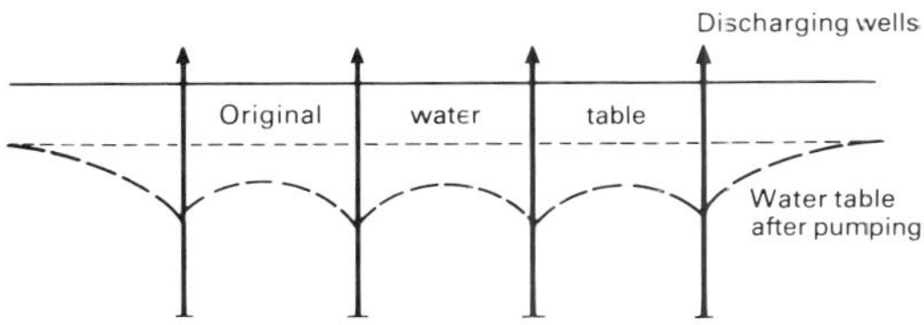

Fig. 7.9 Diagram illustrating well point method of drainage.

excavation, and pumped to produce a cone of depletion in the water table. Once pumping is commenced, it must continue until foundation construction is complete. Soils such as silts and silty clays have low permeability and are therefore unlikely to cause flooding problems. However, within the zone of saturation, excavations in these soils may be quite unstable. Lowering of the water table may increase stability. On the other hand, due to the development of high pore pressures with rapid drawdown, instability may result. It is essential, therefore, that strict control be maintained on rate of drawdown during pumping. Electro-osmosis may be used to lower the water table in less permeable soils. Here perforated casing in bores are designed as electrodes; when an electric field is formed, water is drawn to the cathode, collects in the bore, and is pumped out. Electro-osmosis is also frequently used to stabilize cohesive soils.

Table 7.3 Water quality for various uses. (Concentrations are in p.p.m.)

	Drinking	General household use		Irrigation		Food processing	Boiler water	
		good	poor	good	poor		High pressure	Low pressure
Antimony	0.05					0.05		
Arsenic	0.05					0.05		
Barium	1.00					1.00		
Bicarbonate	500	150	500	200	500	300	5	50
Boron	20			0.3	3.0			
Cadmium	0.01					0.01		
Calcium	200	40	100			80	1	40
Chloride	250			100	300	300		
Chromium	0.05					0.05		
Copper	1.05	0.5	3.0			3.0		
Cyanide	0.2					0.2		
Fluoride	1.5					1.5		
Hydrogen sulphide	1.0	0.05	2.0			0.5	0	5
Iron	1.0	0.2	0.5			0.2		
Lead	0.05					0.05		
Magnesium	125	20	100			40	1	20
Nitrate	20					20		
Phenol	0.001					0.001		
Selenium	0.01					0.01		
Silica		10	50			50	1	30
Silver	0.05					0.05		
Sodium	200	100	300	50	300	300		50
Sulfate	250	100	300	200	500			
Synthetic detergents	0.05	0.2	1.0			0.5	0	0
Total solids	1500	300	2000	500	3000	1000	100	2000
Zinc	5					5		

7.8.2 CONTROL OF GROUNDWATER: GROUTING

Control of the flow of groundwater through a rock mass can be achieved by the injection, under pressure, of some material, or materials, which will reduce the permeability of either, or both, the rock material and rock mass. The technique can also be used to achieve an increase in strength and elasticity of the rock mass. The grout is injected through a standpipe or borehole depending on the type of material being treated and on the grouting pressure. Grouts are of two main types: particulate (cement grout and clay grout) and non-particulate grouts (silicate grouts and organic polymer grouts). Whichever type is used, the purpose is to fill, and seal, all pore spaces and fissure spaces within the rock mass. The type of grout used will almost always be determined by geological conditions.

In the case of a fissured rock mass, the percolation of water occurs mainly along fissures. In grouting such a mass no attempt is made to seal the pores in the rock, since most of the water movement occurs along fissures rather than through pores. The spacing, orientation and condition of the fissures have to be determined as a basis for the design of an efficient grouting programme. In the first place, it is essential to locate, space, and orient the grout holes so as to achieve the maximum intersection of fissures, and thus the most effective penetration of grout. The condition of the fractures is equally important: the more open the fissures, the more grout will be required to seal them effectively. Fissures which are narrower than 100 μm rarely accept grout. The data are also useful in determining the necessary viscosity of the grout mix. Infillings may militate against penetration or acceptance of grout, and the fissures, in such cases, may require flushing out with water before grouting commences. The spacing and openness of the fissures may, to some extent, also determine grout pressures. The stability of the fissured rock mass can be destroyed by grouting at excessive pressures. Geological data can be used to determine whether or not consolidation (low pressure) grouting is necessary prior to high pressure grouting.

Consolidation, low pressure, blanket grouting is usually carried out to shallow depth to provide a stable capping prior to high pressure grouting. Holes are drilled to shallow depth, usually on a square grid, the depth being determined by fissure spacing and width. During this phase, careful observations have to be made to ensure that movement of the fissure-bound surface blocks does not occur, and that grout does not flow at the surface. During high pressure grouting, similar precautions have to be taken. At the conclusion of high pressure grouting, it is desirable that some boreholes be drilled between grout holes, and core recovered for examination by the geologist to assess the effectiveness, and the degree of penetration of the grout. If the core indicates poor penetration, then further grouting will be necessary.

Whereas in fissured rock masses particulate grouts are normally used, grouting of permeable beds or masses of permeable soils usually utilizes non-particulate materials. Permeation grouting using particulate materials is limited by the grain size, porosity and permeability of the soil, and the selection of grouting material is made on the basis of its ability to penetrate the soil. For example, in a coarse sand with high permeability, encountered in a Yallourn power station tunnel, cement grout proved to be highly effective; on the other hand, a fine silty sand refused to accept cement grout and required silicate grouting. In both of these cases, grouting was required not only to control the flow of groundwater, but also to stabilize the soils in the tunnel.

7.8.3 GROUNDWATER IN TUNNELS

One of the major hazards in tunnelling concerns large inflows of groundwater. This can be illustrated by an example from the Kiewa hydroelectric project,

Australia. The No. 4 head race tunnel encountered jointed rock, associated with a fault zone consisting of brecciated gneiss; the actual width of the brecciated zone was about 2 m (Fig. 7.10). Discharges of groundwater of about 60 000 l hr^{-1} occurred, dislodged the closely jointed rock, with total failure of the tunnel at this point. Because of the dangerous working conditions, the decision was made to divert the tunnel around the failure. Investigation drilling showed that the

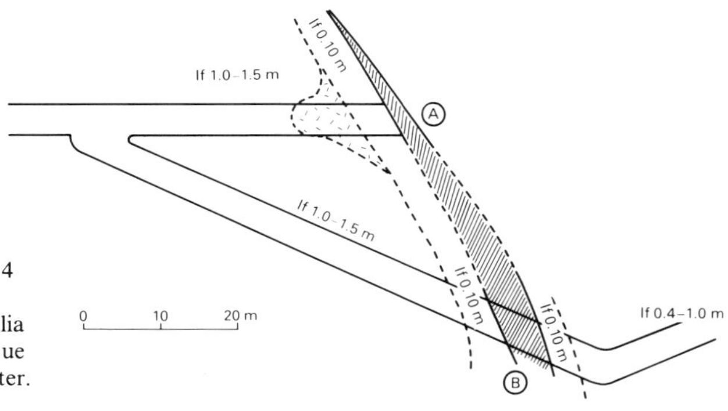

Fig. 7.10 Diversion of No. 4 head race tunnel, Kiewa hydroelectric project, Australia as a result of tunnel failure due to high inflows of groundwater.

brecciated zone was thicker in the proposed new tunnel section, and that high discharges of groundwater were to be expected. In the event, the discharges from the failed section diminished at a regular rate, dropping to about 500 l hr^{-1} after 8 weeks, and, when the new section of tunnel passed through the closely jointed and brecciated zone, only very light timber supports were required, with water inflow less than 100 l hr^{-1}. It is clear that the fractured and brecciated zone constituted an aquifer which during a 10 week period drained sufficiently into the failed section of the tunnel for a cone of depletion to be developed.

External groundwater pressures can be sufficiently great to cause failure by buckling of steel lining in pressure tunnels. When the tunnel is under load, the internal pressure usually balances the external pressure; when the tunnel dewatered, however, inward buckling of the steel lining may occur. These problems are discussed in detail by Jaeger (1955).

7.8.4 GROUNDWATER AND SLOPES

Due to the development of high pore-water pressures, the shearing resistance of the soil constituting a slope may be reduced to a level below the shear stresses acting on the soil. In such a case, failure occurs. If groundwater is a major contributing factor to slope instability, conditions can be considerably improved, in many cases, by lowering of the water table. On slopes, this can usually be achieved using horizontal, or gently inclined bores. At the same time, surface drains should be installed on the slope to reduce infiltration and hence recharge of the groundwater body.

An unusual example is seen in the failure of a cut slope in the Yallourn North brown coal mine, Australia (Fig. 7.11). Following a prolonged period of heavy rain, hydrostatic pressure developed in the sand aquifer confined between the Latrobe Coal Seam and a unit known as the Latrobe Clay, both of which were

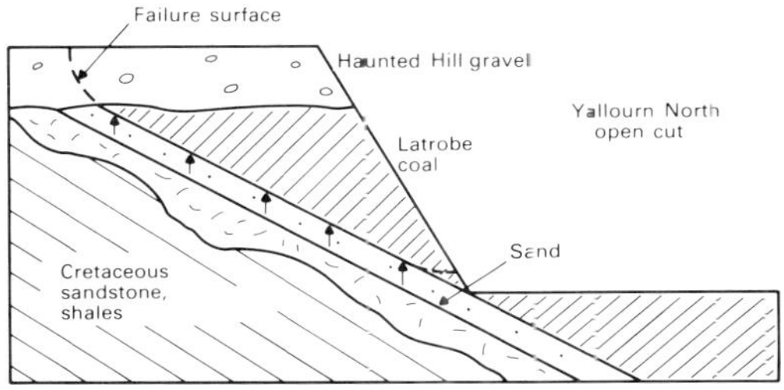

Fig. 7.11 Geological section at site of slope failure, Yallourn North Brown Coal Mine, Victoria, Australia.

impermeable. The water pressure in the aquifer was sufficient to cause the overlying coal to 'float' resulting in the failure of the open cut slope over a length of some hundreds of metres. In those sections of the mine where the geological features were comparable to those at the site of the failure, drainage of the sand was undertaken, and no further failures occurred.

7.8.5 GROUNDWATER AND DAMS

When a dam is constructed, and water is impounded behind the dam, some of the water infiltrates into the surrounding soil and rock, causing a local rise in groundwater level. Under some conditions, serious losses of stored water can occur, either to the groundwater body, or to adjacent catchments. Knowledge of the groundwater regime at and close to a damsite is essential.

Where direct connection exists between the water storage and an aquifer, as is the case at the White Swan Dam, Ballarat, Australia (Fig. 7.12), losses may be

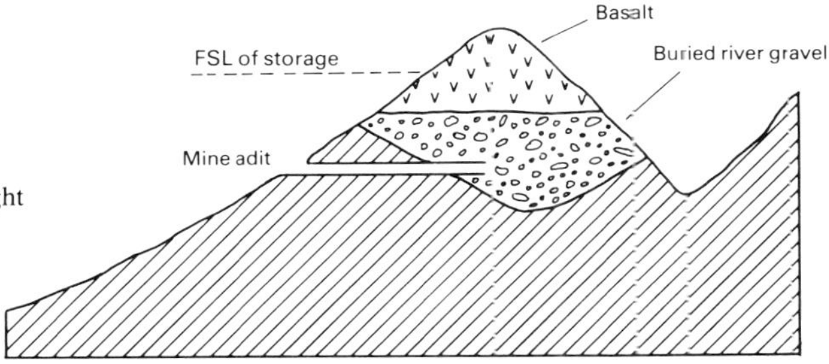

Fig. 7.12 Cross-section of right abutment, White Swan Dam, Ballarat, Victoria, Australia.

too high to be tolerated, and measures have to be taken to prevent access of stored water to the aquifer.

Although not directly involved with groundwater, the problem of piping failure has to be considered. Many examples are known, of course, of piping erosion associated with the interflow. As water flows through a soil, it experiences a loss of head, which is accompanied by the development of a seepage force. If this force acts upwards, it decreases the apparent weight of the particles, and hence the shear strength of the soil. When the upward force becomes equal to the submerged weight of the particles, the soil particles 'float', and the particles act as a liquid. This liquid moves along flow lines causing internal erosion, and a pipe develops. Such piping can lead to failure of an earth dam either within the structure itself, or within the foundation.

While theoretical considerations suggest that piping is not dependent on soil type, practical experience has shown that fine to very fine sands and silty sands of uniform grain size are more susceptible to piping than other soil types. Aitchison and Wood (1965) found that many small earth dams constructed of predominantly clayey soils failed by piping. These workers postulated that post-construction deflocculation of clay is a contributing cause.

References and further reading

Aitchison G. D. & Wood C. C. (1965) Some interactions of compaction, permeability and post-construction deflocculation affecting the probability of failure in small earth dams. In *Proc. 6th Int. Congr. Soil Mechs.* Canada, pp. 442–6.

Bell F. G. (1978) Petrographical factors relating to porosity and permeability in the Fell Sandstone. *Quart. Jl. Engng. Geol.* **11**, 113–26.

Bouwer H. (1978) *Groundwater Hydrology*. McGraw-Hill-Kogakusha, Tokyo.

Davis S. N. & De Wiest R. J. M. (1966) *Hydrogeology*. John Wiley, New York.

Heath R. C & Tramor F. W. (1968) *Introduction to Groundwater Hydrology*. John Wiley, New York.

Jaeger C. (1955) Present trends in the design of pressure tunns. In *Proc. Inst. Civ. Engs.* pp. 545–96.

Pless T. (1975) Permeability of some rocks from the Sydney Basin. Unpublished MAppSc thesis, University of New South Wales.

Rzhevsky V. & Novik G. (1971) *The Physics of Rocks*. MIR, Moscow.

8

Geological Investigations

8.1 Introduction

At the site proposed for almost every civil engineering structure it is essential, for safe and economic design and construction, that the geological conditions and the engineering properties of the soils and rocks be known. Even at the initial planning stage, some knowledge of geological conditions is essential, and the more advanced the stages of planning, design and construction become, the more detailed becomes the geological data required. Geological investigations begin at the planning stage, and continue throughout the project until foundation construction is complete.

Geological investigation involves geological mapping at scales which are determined by the stage of the investigation; subsurface investigation by means of bores, test pits and cuts; geophysical surveys; and field and laboratory tests to determine the engineering properties of the rocks and soils. Geological investigations for civil engineering purposes require a specialized approach, quite different from that for, say, a mineral resources investigation. It is essential that the geologist be specially trained to carry out engineering site investigations.

8.2 Terrain evaluation

Engineering geology and terrain evaluation are complementary disciplines; the former is concerned with the collection of geological data for engineering purposes, the latter provides methods for processing the information, and is an essential part of any engineering geological investigation. The terrain, as it exists at present, is the result of the interaction between geology and climate with time. This is the basic principle underlying terrain evaluation. Many systems of terrain evaluation have been proposed for a variety of purposes, including engineering. A system most applicable to engineering is one based on easily recognizable natural features which reflect geomorphic attributes. The scheme now generally used is that due to Aitchison and Grant (1967). The basis of this is that any area of land can be defined uniquely in terms of its slope characteristics, soils, vegetation and underlying geology (lithology and structure). This scheme, known as Pattern-Unit-Component Evaluation (PUCE) recognizes four classes: terrain component, terrain unit, terrain pattern, and province.

Table 8.1 summarizes the methodology of this system of terrain classification. It will be noted that the classification relies on both remote sensing and field data, the latter becoming more important as the degree of generalization is reduced, and the study becomes more detailed. The 'province' is a class based almost exclusively on remote sensing data, and forms a basis for regional planning. At the other extreme, the 'terrain component' is defined by data collected almost exclusively in the field, and forms the basis for detailed site investigations.

8.2.1 TERRAIN COMPONENT

Apart from microtopographic features, a terrain component has a constant rate of change of slope, which may be concave, or convex, or uniform, but not concavo-convex. The component has a consistent association of soils and vegetation which reflect a uniform geology. Grant (1971) refers to major and minor axes of slopes; the former is that line joining the highest and lowest points of a slope, and the latter is a line at right angles to this, and slope is measured along these axes. Such features as microtopography and rock outcrops are not taken into account in classification. However, it is essential that they be considered in description and in the separation of terrain components. For example, an area which but for microfeatures such as gilgai in one part and rock outcrops in another would be regarded as a single unit, would require division into two components, since the engineering characteristics of gilgai soils are quite different from soils lacking this microfeature. Again, an area with rock outcrops differs, in engineering terms, from one which lacks such outcrops.

The exclusion of such features as gilgai, rock outcrops and other microtopographics from the classification is a practical issue. In order to have a workable scheme, a limit has to be placed on the variation permitted. This may be a matter of subjective choice.

8.2.2 TERRAIN UNIT

A 'terrain unit' consists of a number of 'terrain components' which together make up a single geomorphic feature. The characteristic slope, soil and vegetation associations of the terrain unit will be those produced by a synthesis of the constituent terrain components. The terrain unit can be divided into classes, each class having a characteristic association of slopes and a locally consistent relief:

Class A

Flat or gently undulating surfaces. These surfaces are not necessarily part of the one erosional or depositional geomorphic surface.

Class B

Slopes between Class A surfaces.
These slopes may be: (a) gentle; (b) steep; (c) escarpments.

Class C

Isolated hills and ridges, but excluding flat topped hills such as mesas, which would fall in Class A (surface) and Class B (slope).

Table 8.1 Terrain classification methodology.

Initial source of data	Intermediate stage in classification: Relevant factor in classification	Intermediate stage in classification: Title	Supplementary source of information	Stage in terrain classification: Relevant factor in classification	Stage in terrain classification: Title	Stage in terrain classification: Mode of expression
Geological maps or aerial photographs on scale of the order of 1 : 10^6				Areas of constant geology at the stratigraphic group level Areas of similar remote sensing pattern	**Province**	Map 1 : 250 000
Aerial photographs on scale of the order of 1 : 10^4	Areas of similar airphoto pattern	Aerial photography pattern	Ground study of geomorphology and association of terrain units	Areas of similar airphoto pattern + constant geomorphology + constant association of terrain units	**Terrain pattern**	Map 1 : 250 000 Statement of association of terrain units
Photogrammetric studies of aerial photographs on scale of the order of 1 : 10^4	Single geomorphic feature	Landform unit	Ground study for recognition and assessment of dimensions of physiographic features (where not obtained from aerial photography studies): ground study of association of earthen materials and vegetative cover: studies of association of terrain components	Areas occupied by single physiographic feature with characteristic association of earthen materials and vegetation: areas with similar associations of terrain terrain components	**Terrain unit**	Map 1 : 50 000 Statement of characteristic combination of topography, and association of vegetation and earthen materials; and of relative dominance within terrain pattern
			Ground studies for recognition of specific slopes, soil, surface cover, and vegetation associations: detailed geological field studies	Areas of constant rate of change of slope: constant soil at primary profile level: consistent vegetation associations; geological features	**Terrain component**	Statement of characteristic combination of slopes, vegetation and soil; and of relative dominance within terrain unit. Mapped only in special cases.

Table 8.2 Typical description of a terrain pattern in terms of terrain units. Lithology: siltstone, claystone, sandstone, minor conglomerate. Siluro-Devonian, tightly folded, jointed, deeply weathered. Occurrence: widespread in areas underlain by Silurian-Devonian rocks; areas of lower relief. Topography: undulating terrain containing smooth sloped isolated hills. Terrain units: as shown. [After Grant (1968).]

Number	Terrain pattern area (%)	Occurrence	Topography	Description of dominant: Soil	Land use	Vegetation
1	5	Discontinuous; floodplains adjacent to drainage channels	Flat surface	Uniform light grey or grey-brown deep clayey silt	Melbourne area, parkland; other areas pasture, some agriculture	Mostly cleared, grassland, scattered trees, messmate, manna gum or cultivated.
2	80	Continuous; extensive	Undulating surface. Crests: rounded, linear to 5° Slopes: convex to 10°	Duplex light grey clayey silt to 20 cm over yellow mottled heavy-textured clay over CW rock at variable depth	Melbourne area, urban development; other areas pasture, some agriculture	Built up or mostly cleared, grassland, scattered trees; some areas woodland.
3	5	Sometimes leading to drainage channels	Dissected slope. Slopes: convex to 10°	Duplex light grey clayer silt to 20 cm over yellow mottled heavy-textured clay over CW rock at variable depth	Melbourne area, urban development; other areas pasture, some agriculture	Built up or mostly cleared, grassland, short grass, scattered trees; some areas woodland.
4	10	Inclusive in undulating surface	Complex hill. Crests: rounded, convex to 10°	Shallow duplex light grey or grey-brown clayey silt to 10 cm over yellow mottled heavy-textured clay to 1 m or Crests and upper slopes: light grey or grey-brown clayey silt to 10 cm over decomposed rock with some rock outcrop	Melbourne area, urban development; other areas pasture	Built up or mostly cleared, grassland, short grass, scattered trees; some areas woodland.

				Lower slopes: duplex light grey clayey silt to 20 cm over yellow mottled heavy-textured clay to 1 m over decomposed rock		
5	5	Inclusive in undulating surface	Elongated rounded hills Crests: rounded, convex to 5° Slopes: convex to 10°	Shallow duplex light grey or grey-brown clayey silt to 10 cm over yellow mottled heavy-textured clay to 1 m or Crests and upper slopes: light grey or grey-brown clayey silt to 10 cm over CW rock with some rock outcrop Lower slopes: duplex light grey clayey silt to 20 cm over yellow mottled heavy-textured clay to 1 m over CW rock	Melbourne area, urban development; other areas pasture	Built up or mostly cleared, grassland, short grass, scattered trees; some areas woodland.
6	<1	Drainage often artifically covered and controlled in the Melbourne area	Minor stream channel, often depressional. Slopes: concave to 10°	Uniform or stratified dark brown, grey-brown or occasionally yellow mottled silty or clayey fine sand or silty clay, some angular gravel towards stream heads	Unused	Grassland, short grass, manna gum, peppermint, long leafed box.

Class D

Water courses and lakes.

Within the terrain unit, the geology is constant. However, gradations in soil types will occur due to varying geomorphic conditions. For example, the soils on surfaces will have different textural and profile characteristics to those on slopes. Similarly, gradational variations in vegetation are to be expected.

8.2.3 TERRAIN PATTERN

A 'terrain pattern' is an area of constant geomorphic association consisting of recurring topographic, soil, and vegetational associations. The terrain pattern is, in effect, made up of a limited number of terrain units. Boundaries of patterns are fixed by marked changes in terrain units, changes in the type and density of drainage channels, and marked changes in relief amplitude.
Table 8.2 is a simplified version of part of a terrain evaluation, at the pattern level, of an area near Melbourne, Victoria. This emphasizes the importance of the topographic classes in relation to terrain units, and the variation in soils within a given unit or component.

8.2.4 PROVINCE

A 'province' is an area of constant geology at the level of the stratigraphic group, or of a large pluton or lava field. Topographic soils and vegetational features will vary considerably throughout the province. The province is clearly constituted by a number of terrain patterns, some at least of which may be recurring, depending on the geological structure and the regional geomorphology.

8.2.5 TERRAIN EVALUATION AND ENGINEERING GEOLOGICAL INVESTIGATIONS

Terrain evaluation does not replace site investigation, but rather supplements it by providing a rational basis. Engineering geological investigations may be carried out at two levels: on a regional basis to assist in the selection of the most satisfactory location for a project; and as a detailed site investigation. Classification to terrain pattern level (or, in exceptional cases, terrain unit level) is usually sufficient for regional studies. However, for detailed site investigation, classification to the terrain component level is essential.

The area described by Table 8.2 served as a basis for site and material investigations for a new highway, although the original purpose of the evaluation was for urban and regional planning. When the terrain has been classified, engineering geological and geotechnical information can be assigned to appropriate members of each class of terrain. Since a basic assumption of terrain evaluation is that all occurrences of each member of each class are homogeneous, engineering geological and geotechnical data applicable to one terrain class member can be extrapolated to all occurrences of that member. On this basis, it is possible to predict geotechnical characteristics and material sources over the area evaluated and classified.

8.3 Geological maps

The most concise and clear method of presenting geological data is on a geological map. Such maps are prepared at a variety of scales for a variety of purposes. For example, a map to depict, simply, the geology of a region in terms of rock types, their distribution, relative ages, and structure may be at a scale of 1:25000. On the other hand, a map showing quite detailed information for a mineral prospect or an engineering site may be at a scale as large as 1 : 500, or even larger. It must be remembered that the detail, accuracy and method of mapping will all change with the scale at which the mapping is carried out. For example, for small scale maps, the procedure involves the interpretation of aerial photographs with field observations recorded directly on these photographs. Major lithological units and major structures only are recorded. Where detailed, large scale maps are required instrumental surveying by tachometry is used, while every possible geological feature is recorded. Engineers should be aware of the range of reliability, detail and accuracy; they should also be aware that, the smaller the scale, often the greater the amount of interpretative, as opposed to factual, content of the map.

Geological maps prepared for civil engineering purposes record not only basic geological information, but additionally, geological data of value to the engineer for planning, design and construction purposes. Engineering geological maps themselves vary depending on the purpose of the map. Students are referred to the UNESCO publication *Engineering Geological Maps* for a full presentation of the various types of such maps.

The geological map is a two-dimensional representation of three-dimensional data. From his mapping, the geologist is able to predict, from surface observations, subsurface geology. The predictions, however, must be checked by methods other than direct surface observation if the information to be given to the engineer is to be totally reliable. Some subsurface conditions, e.g. weathering, can be predicted from surface observations in such a general way that they are of value only at the planning stage of an engineering project. In order to resolve doubtful geological relationships and conditions, it is essential, in engineering geological investigations, for some subsurface exploration to be carried out.

The engineering geological maps for an engineering project should show the following:

(1) Geomorphological features, including unstable areas, and areas of erosion and deposition.

(2) The distribution of rock and soil types, their character, stratigraphical relationships, structural features, and physical and engineering properties.

(3) Weathering conditions, including thickness of weathered zone, grades of weathering, and properties of weathered rocks.

(4) Hydrogeological conditions, including water table and fluctuations, aquifer properties, springs, and groundwater quality.
An example of a basic geological map prepared for a site investigation is shown on Fig. 8.1.

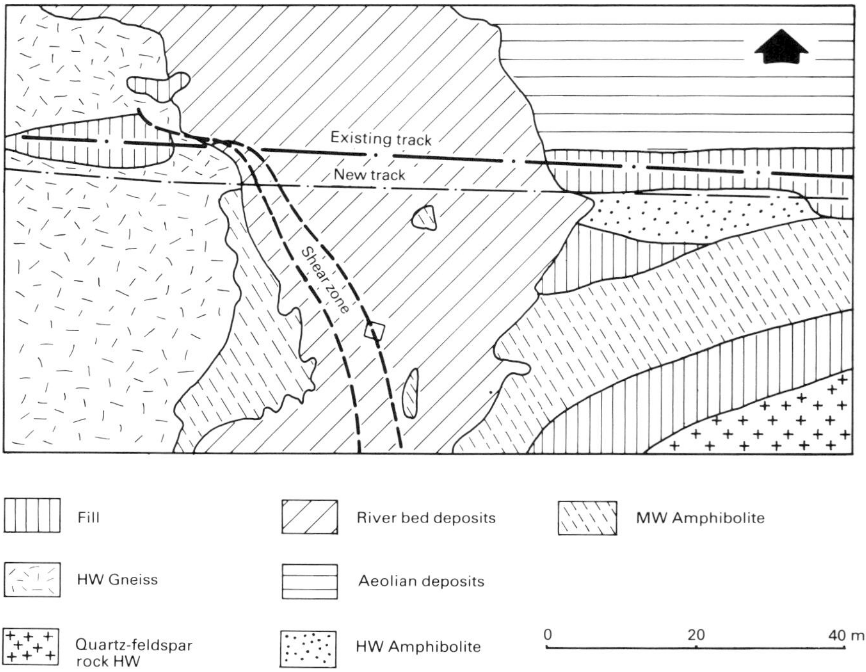

Fig. 8.1 Engineering geological map prepared for site of new railway bridge, Broken Hill, Australia.

8.4 Subsurface investigations

Investigations of subsurface geological conditions are of two main types: direct and indirect. The former involve direct observation of the rocks and soils, the latter, the interpretation of geological conditions based on the measurement of certain physical properties. Direct methods include drilling and boring, shaft sinking, trenching and tunnelling. Indirect methods utilize geophysical techniques. For the detailed geological investigation of an engineering site it is standard practice to use both direct and indirect methods.

8.4.1 DRILLING

Rotary drilling and percussion boring are the two groups of methods used for direct subsurface investigation and sampling. Rotary methods include auger drilling, diamond drilling, and calyx drilling.

The first of these methods uses a hand operated or machine driven auger, and produces a disturbed sample. Its use is limited to soils and unconsolidated sediments, while the depth to which drilling may be carried out is relatively shallow. The method is ideal for shallow investigations for foundations in soils, unconsolidated sediments, and completely weathered rocks.

Diamond drilling is the technique most frequently used for deep investigation of rock. In theory, a continuous core is produced, which can be examined to determine rock type, condition of the rock, including grade of weathering, the type, spacing and condition of fractures, rock material and rock mass quality and evidence of major structural features such as faults and folds. Samples are available for testing of properties. The borehole itself provides hydrogeological data, while television and photographic techniques are available to examine the walls of the borehole.

The success of diamond drilling depends to a large degree on the type of equipment used, and the skill of the operator. Core is cut by a bit into which are set industrial diamonds. The core passes up into a core barrel, from which it is extracted when the length drilled reaches the length of the core barrel, or earlier, if the operator experiences drilling conditions which require him temporarily to cease drilling. The type of core barrel used is critical. So-called single tube core barrels, which consist of a single steel tube, should never be used. Core in such a barrel is subjected to vibration, rotation of the barrel, and scouring by drilling water which passes down the barrel. As a result, all but the strongest core will be lost, and it is frequently the weaker core which is of most interest to the engineer.

The type of core barrel most commonly used at the present time is the M series triple-tubed core barrel with split innermost tube. Because of the arrangement of the tubes, core in such core barrels is free from effects of rotation of the outer tube, and from the scouring action of drilling water, while the effects of vibration are very much reduced. The correct selection of bit size is also an important factor in ensuring high core recovery of even the softest rock. The standard sizes used for engineering geological investigations are EX (18 mm diameter); AX (30 mm); BX (41 mm) and NX (54 mm). The actual cores recovered have slightly smaller diameter, e.g. NX core has diameter of 47 mm. It is usual practice to begin drilling with the largest size bit, and progressively to change to smaller bits as the quality of the rock improves below the zone of weathering.

The logging of core for engineering purposes requires great care and skill, both in observing and recording. Standard procedures are set out in the Geological Society Engineering Group Working Party Report (1970). A sample log is shown in Fig. 8.2. The log should record details of drilling techniques and drilling progress, with particular mention being made of loss or return of drilling water, and the static level of water in the bore. Geological aspects to be recorded are the core recovery, and RQD, and fracture spacing index, for each run; rock type and condition (including weathering grade); structural features including type and condition, and estimates of strength (usually point load index). As much information as possible should be recorded graphically.

Even the most carefully logged core should not be destroyed, but it should be carefully stored for future reference at least until the project for which the drilling was done has been completed. During design and construction it is not uncommon for reference to be made to core.

Diamond drilling has an advantage over most other techniques not only in the production of an 'undisturbed' core, but also in the fact that the direction of drilling can be varied between vertical and horizontal. It is often technically desirable, and more economical, to drill inclined or horizontal holes. This will be discussed later in this chapter. One problem develops, however, in that inclined and horizontal drill holes frequently deviate from their initial direction and inclination. For this reason, if the data obtained from the core are to be reliable, it is essential to survey all inclined and horizontal boreholes for both direction and inclination.

The third main method of rotary drilling used in engineering investigations is the calyx technique. In this case, a rotating steel bit combined with chilled steel shot serves as the cutting agent. Calyx drills are of various sizes, but the two most common are 48 in. and 72 in. diameter. The main value of these drills lies in their size, since the geologist and engineer can be lowered down the borehole and examine the rock *in situ*. However, the cost of calyx drilling large diameter holes is extremely high, and the technique is used only in exceptional circumstances.

Percussion drilling depends on the breaking up of the rock or soil by impact, and sampling at regular intervals with a sampling tube. The method is suitable only for soils and soft rocks, and suffers the further disadvantage that the samples obtained are broken and disturbed. A more serious, and sometimes critical disadvantage of the method is that it may fail to disclose relatively thin, but possibly highly significant beds or structures. For example, the foundations for the Yallourn D Power Station in the Latrobe Valley, Victoria, were investigated using percussion drilling. Samples of brown coal were correctly described, but the broken nature of the samples in a saturated condition precluded the recognition of a layer of thixotropic weathered coal up to 50 cm which developed on the fresh coal. It was not until foundation construction was in progress that the existence of this low strength, thixotropic material, was recognized. As a result, foundation design required considerable modification. Similarly in the Fisherman's Bend Industrial Development, Melbourne, percussion drilling did not indicate a continuous bed of black, organic, highly compressible silt 20 cm thick, which again was only discovered during foundation construction and which also caused design alterations.

Although the method is used in engineering investigations, it finds its greatest application in water boring. For this purpose, fine geological information is usually not required; the main purpose of the bore is to serve as a basis for the production of groundwater, and geological information gathered is used primarily to define relatively thick aquifers and aquicludes. Even so, in this case, percussion drilling provides no information relating to structural defects, such as joints, which may be very important in hydrogeological studies.

Reference is sometimes made to wash or jet boring for engineering geological investigations. The method is based on using a water jet to wash out earth materials which are then examined. Extensive experience has indicated that not

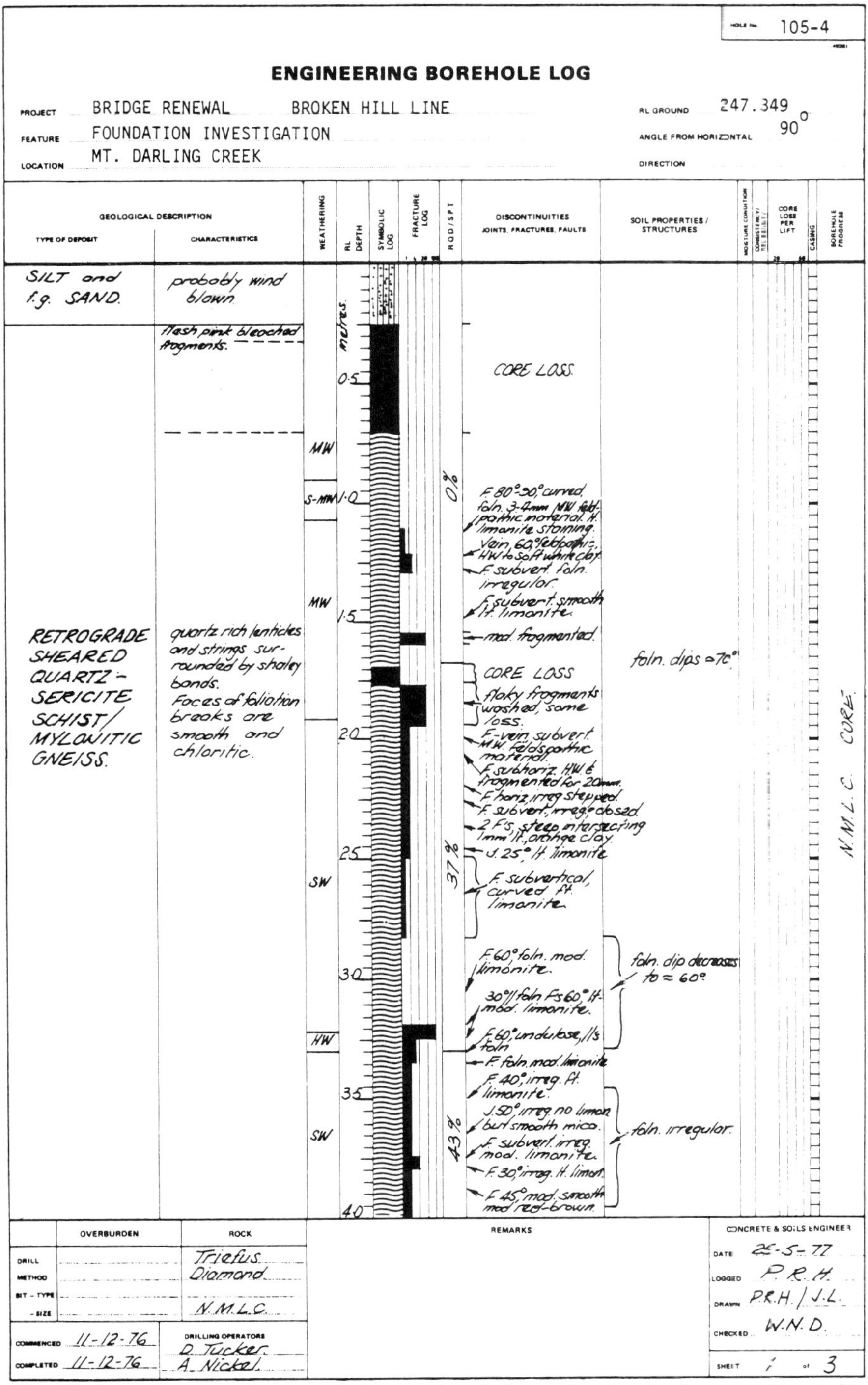

Fig. 8.2 Part of the log of a bore drilled to test foundation rock at the site of a new railway bridge, Broken Hill, Australia (Davies 1980).

only is this technique of no value for engineering investigations, but also that it gives erroneous results. The method should never be used for site investigation purposes.

Although core or other samples are obtained from drilling, *in situ* conditions are *interpreted* from these samples. Experienced engineering geologists are fully aware that the most careful interpretation may not always present an accurate description of the actual geological conditions. Modern techniques using borehole cameras and borehole television permit the geologist to examine the walls of the borehole and to combine the data so obtained with the data from the core or other samples. In this way, the accuracy of the interpretation is increased considerably. For large diameter calyx drill holes, the wall of the hole can actually be mapped.

Boreholes of any considerable length, and especially inclined or horizontal diamond drill holes, do not maintain the design orientation, but can gradually diverge from the set out direction and inclination. If the geological information obtained from the drilling is to be of full value, it is essential to know the precise position of the borehole. For this reason, it is essential that all horizontal and inclined boreholes be surveyed for direction and inclination. Instruments are available for this purpose, recording photographically the direction and inclination along the borehole at predetermined intervals. Since determination of direction is based on the use of a magnetic compass, steel casing, or the presence of magnetic minerals in the rock will give erroneous readings. As technology develops, this serious problem may be overcome.

A drilling programme for engineering geological investigations can be very costly. It is therefore essential that the programme be very thoughtfully planned to give the maximum of information for the minimum cost. Planning is a joint exercise of geologist and engineer, and is controlled by the geological problems which require to be solved to provide the engineer with the information he requires. The site engineering geological maps should serve as a basis for deciding the number of bores required, their location, direction, inclination and depth. If structural features are essentially horizontal, vertical drillholes will provide the most information at minimum length of hole. If, on the other hand, the structural entities are vertical, then horizontal holes would be the most efficient. For inclined structures, boreholes should be drilled as near normal to the dip as is practicable (Fig. 8.3). Careful design of a drilling programme can reduce costs of the total investigation quite significantly.

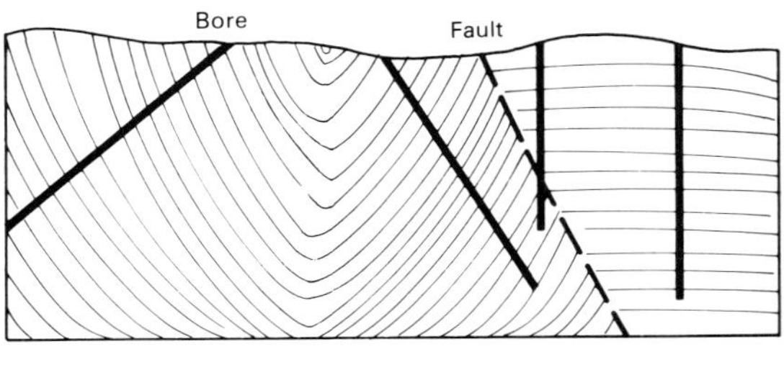

Fig. 8.3 Drilling of boreholes in relation to structure to achieve maximum information.

8.4.2 GEOPHYSICAL INVESTIGATIONS

Geophysical methods measure some specific property of the rock mass which is interpreted in geological terms. Most methods are available for engineering investigations, but only two or three are generally used. Seismic techniques, which measure the velocity of sound in rock; and electrical resistivity techniques, which measure the electrical resistivity of the rock are the two methods most applicable. Radioactive methods are used in borehole logging. Magnetic and gravity techniques have only limited value for the engineering geologist.

Seismic techniques used are of two types: seismic refraction, which is the method generally used on land; and seismic reflection, used in the solution of marine problems. Seismic methods require a velocity contrast to exist in the rock mass. The seismic refraction method depends on measuring the velocity of sound waves in rock. Sound waves travel from the source down to the interface between two rock units with differing velocities. At the interface, the wave is refracted, and its time of arrival at the recording station is determined. Time–distance graphs (Fig. 8.4) are prepared from which the thickness,

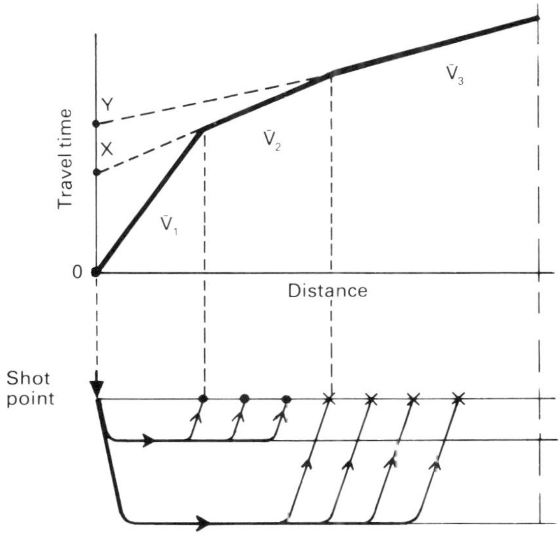

Fig. 8.4 Seismic refraction time–distance graphs.

orientation and seismic velocity of a layer can be determined. Correct geological interpretation requires some additional data, usually provided by a drillhole. Seismic reflection profiles provide travel time data, and record a structural profile.

Electrical resistivity methods are most useful in areas where the geological units have strongly contrasting electrical properties, usually governed by porosity and moisture content. These methods are therefore particularly useful for delineating sands, gravels and clays. Resistivity profiling has been used, with some success, in defining fault zones in massive igneous rocks. Resistivity logging of boreholes is used to give additional information which can supplement core data.

The advantage of geophysical techniques lies in the rapidity, and relatively low cost, with which information is obtained. The site of the McKay underground power station, Australia, was on deeply weathered granodiorite, and the precise location depended on the weathering profile, and the location of two thick fault zones. Seismic refraction surveys gave data, after 3 weeks, which it was estimated would have required some 4 months of diamond drilling, at 12% of the estimated cost of the drilling. Zones of completely weathered, moderately weathered and fresh rock were defined, while the approximate positions of the fault zones in the dense scrubby terrain were delineated (Fig. 8.5), permitting the

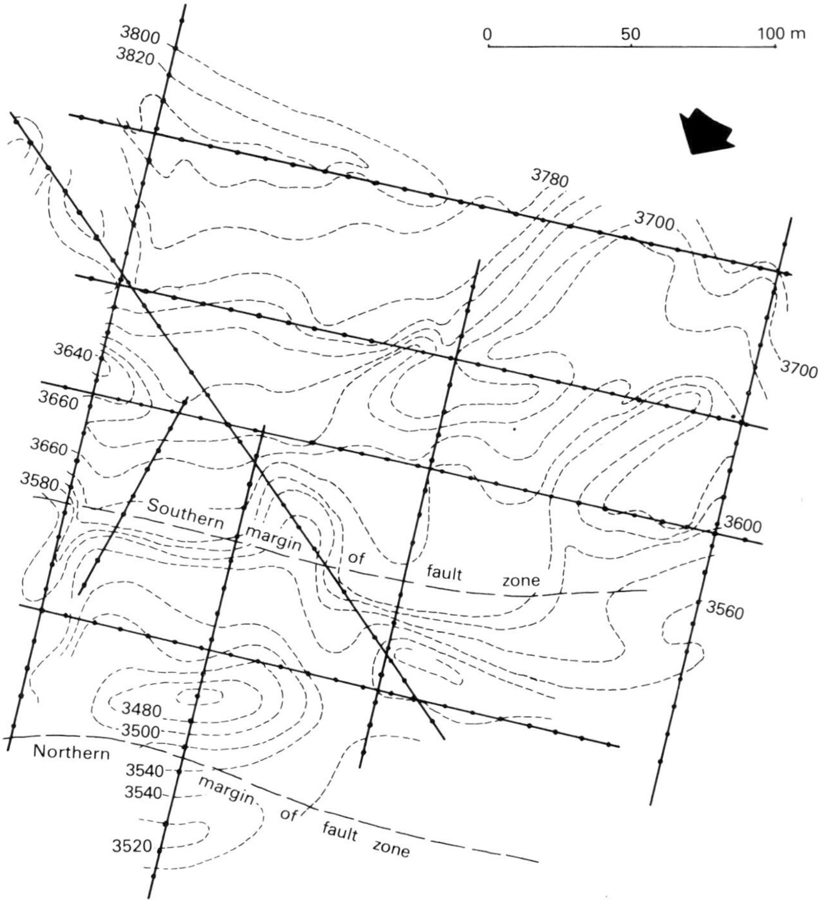

Fig. 8.5 Contours of fresh rock surface determined by seismic refraction at McKay power station site, Kiewa, Australia.

design of an efficient diamond drilling programme to investigate the engineering characteristics of the faults.

Despite the value of geophysical methods of investigation for engineering works, a number of problems may arise. For example, in seismic refraction surveys difficulties may arise in the identification of a low velocity layer underlying a high velocity layer. This problem occurred during tunnel investigations in Tasmania, where high velocity dolerites overlie low velocity sedimentary rocks. In the case of resistivity studies, a horizontal surface, and stratification are assumed. In an irregular topography, resistivity measurements become difficult to interpret. Furthermore, since moisture content of the rock influences resistivity, the depth of the water table will influence the results, and where the water table is irregular, interpretation of the results may be quite difficult.

8.5 Other methods of investigation

So-called wash borings are sometimes used for engineering geological investigations of soils, or unconsolidated sediments. Most engineering geologists believe that these are of no value, in fact, may give misleading data, unless the boreholes are subsequently logged using geophysical methods such as gamma ray logging and resistivity logging.

It is sometimes desirable that more precise information is required than is provided by test bores and geophysical surveys, or that *in situ* tests be conducted. For either of these reasons, test shafts, test trenches, or test tunnels may be excavated. From an investigational point of view, such excavations have the advantage that the rock or soil can be examined *in situ* and the behaviour, in excavation, studied. The disadvantage is the high cost of the work. For this reason, excavations such as shafts and tunnels are usually planned to serve some other purpose, as well as their prime purpose, in the construction phase of the project.

To illustrate this point, reference may be made to the investigation for the McKay underground power station, Victoria, Australia. Here, apart from seismic refraction surveys and diamond drilling, an exploratory shaft was sunk, and an exploratory tunnel driven to examine *in situ* the well developed and closely spaced sheeting which, it was believed, could influence stability of the station arch. Extremely valuable engineering geological data, apart from that directly sought, were obtained. Subsequently, the shaft was broken out to full size to serve as the access to the station, and the tunnel broken out to full size to serve as the tail race tunnel.

All exploratory excavations must be carefully surveyed geologically, and accurate (including photographic) records prepared. These will indicate rock types, condition of the rocks, structural defects, and groundwater conditions. In addition, comments relating to the stability of the excavations, any problems encountered, and the location and type of support required should be included in the records.

8.6 Laboratory investigations

Field investigation of an engineering site is, in itself, inadequate for design purposes. It is essential that laboratory testing of materials be performed as an integral part of the site study. The tests carried out will depend on the project being investigated, and the nature of the site. In general, for soils, tests are made to determine grading, index properties, density, compaction, consolidation, and settlement characteristics, strength and permeability. Under some circumstances, special tests may be required, e.g. for dispersion, stabilization, and clay mineralogy and chemistry.

For rocks, strength and deformation (elasticity and creep) tests, shear parameters, durability and permeability tests are basic. If the rock is to be used as a road aggregate or a concrete aggregate, the rock will need to be tested to determine its suitability for the purpose.

In the laboratory phase of the investigation, the engineering geologist will undertake geometric and kinematic analysis of fracture systems, which may require laboratory testing of the shearing resistance on the fractures, using the Hoek shear box.

All test results obtained in the laboratory require careful analysis by both the geologist and the engineer. In particular, consideration must be given to the applicability of the results of tests on soil and rock materials to the soil or rock mass. If possible, relationships should be established between field- and laboratory-determined properties. For example, the deformation modulus values from plate loading tests in the field are commonly up to one order of magnitude lower than those measured in the laboratory. It was found by Evans (1978), for sandstones in the Sydney Basin, Australia, that a reduction index could be established which would relate values of the modulus of elasticity determined from laboratory tests to field values. It was found too that this index was an index of the frequency of joints in the rock mass. At least for the strength and elasticity of rocks, the laboratory tests give a much higher value than the *in situ* value, and this fact has to be borne in mind by the designer using test results.

8.7 Reporting geological investigations

The purpose of the geological investigation is to present to the engineer geological data which is required for planning, design and construction. The presentation is in the form of a report which, because of its purposes, differs in essentials from a standard geological report. The engineer's prime interest lies in those geological factors which influence safe and economic design and construction. For this reason, the geologist's report will concentrate on this area. Nonetheless, engineers should be aware that proper comprehension of these special aspects is dependent on an appreciation of the geological features at the site. The report, therefore, will begin with a description of the rock types, their distribution, ages and relationships, and a concise statement on structure and weathering. This section of the report will be illustrated by geological maps and sections, and structural diagrams.

The second section of the report will consider the geological factors of relevance to the engineer. These will, of course, vary with the nature of the project, but will almost invariably include detail of structural features and weathering, accompanied by the appropriate engineering geological and geotechnical maps. Engineers will not expect engineering aspects to be considered, but the geologist should draw attention to those features which it is considered may be of importance, e.g. faults, joints and bedding planes, in so far as these could influence slope stability, foundation stability, or leakage.

The third section of the report will present the results, and interpretation, of field and laboratory tests, drawing attention to the relationship between properties determined in the laboratory, and the field values of these properties. The sources of data, such as details of bores, geophysical surveys and laboratory tests should be included as appendices. The report will end with a summary of the conclusions reached, and recommendations for further investigation.

The presentation of the investigation report does not constitute the end of the engineering geologist's work. Rather, his observations will continue throughout the construction period, and it is almost certain that geological features will be disclosed which demand some changes to design. In any case, a detailed record is required of geological conditions at the site for future reference On completion of the project, a final geological report should be prepared which states clearly the geological factors which proved to be of engineering significance, as well as including an objective comparison of the geological predictions, and the conditions actually encountered. In this way, a case history is recorded which serves as a contribution to knowledge in engineering geology.

References and further reading

Aitchison G. D. & Grant K. (1967) The PUCE programme of terrain description. In *Fourth Regional Conference on Soil Mechanics and Foundation Engineering*. Cape Town.

Anon. (1982) *Drillers' Training and Reference Manual*. National Waterwell and Drilling Association of Australia, St. Ives.

Anon. (1976) Engineering Geological Maps. UNESCO, Paris.

Cratchley C. R., Grainer P., McCann D. M. & Smith D. I. Some applications of geophysical techniques in engineering geology. In *Proc. 24th Int. Geol. Congr.* pp. 163–75.

Davies W. N. (1980) Geological investigation for bridge sites near Broken Hill. Unpublished MAppSc thesis, University of New South Wales.

Edwards R. J. G. (1972) The engineering geologist in project reconnaissance and feasibility studies. *Quart. Jl. Engng. Geol.* **4**, 283.

Evans R. S. (1978) Time dependent factors influencing rock slope stability, unpublished Ph D thesis, University of New South Wales, Sydney, Australia.

Geological Society Engineering Group Working Party (1970) The logging of rock cores for engineering purposes. *Quart. Jl. Engng. Geol.* **3**, 1.

Grant K. (1968) A terrain evaluation system for engineering. CSIRO, Div. Soil Mech. Tech. Paper 2, 1968.

Grant K. (1971) Terrain evaluation as a basis for engineering geology. In *Proc. Aust. N.Z. Conference on Geomechanics, Melbourne*.

Griffiths D. H. & King R. F. (1965) *Applied Geophysics for Engineers and Geologists*. Pergamon Press, Oxford.

Jakosky J. J. (1950) *Exploration Geophysics*. Trija, Los Angeles.

Knill J. L. & Price D. G. (1972) Seismic evaluation of rock masses. In *Proc. 24th Int. geol. Congr.* pp. 176–82.

9

Geological Factors in Engineering Works

9.1 Introduction

In the preceding chapters, the basic principles of geology and the general application of these principles to civil engineering have been explained. In the material which follows, specific applications of geology to engineering works are discussed, illustrated by actual examples, or case histories. It is from the experience of actual construction and service that the fundamental issues of engineering geology have been established. Although both engineers and engineering geologists learn from mistakes, i.e. failures of engineering structures, they also learn from successes. The case histories include both failures and successes, deliberately selected to illustrate principles.

9.2 Dams

Dams are structures designed, and constructed, to impound water in a reservoir. The dams may be of concrete, or of earth and rock. Concrete dams are of three basic types: (i) mass gravity, in which the load of, and on, the dam is spread over a large area, and therefore one in which the foundation pressures are relatively low; (ii) buttress, in which the load is carried to the foundation through a series of narrow buttresses, and hence foundation pressures are very high; and (iii) arch dams, which are thin structures, and again, foundation pressures are high. This is particularly true in the abutments where pressures are transmitted to the foundation rocks by arch action. Foundation pressures in the channel section can be quite low. It can be inferred from these statements that the mass gravity dam is a solution for those sites in which the rock foundations have less than superior strength, although relatively high shear and compressive strength may be required in the foundation rock. Buttress and arch dams require high quality foundation rocks. Earth dams can be, and often are, founded on soils, but frequently, particularly if the soil is of no great thickness, the core is founded on rock. Other factors, have to be taken into account. At a damsite, the basic geological requirements are:

(i) suitable foundation materials, in terms of strength and deformation characteristics; (ii) durable rock; (iii) structural patterns such that the foundations are stable and leakage along structures is minimal; (iv) for concrete dams, shallow weathering in order to avoid excessive excavation of weathered material; and (v) stable slopes.

9.2.1 ST FRANCIS DAM, CALIFORNIA

The experience of the St Francis Dam demonstrates that no geological feature should be overlooked. The dam, completed in 1926, was a mass gravity concrete structure, 62.5 m high and 213 m long, constructed on the San Francisquito Canyon. The south-eastern abutment, channel section, and lower part of the north-western abutment of the dam were founded on mica schist, which is faulted against a younger conglomerate. This conglomerate constituted the foundation of the north-western abutment. The foliation in the schistosity dipped roughly parallel to the south-eastern wall of the valley. Zones of shearing were numerous. The rock had low strength and disintegrated readily. The conglomerate was a coarse textured rock with abundant gypsum. During construction, the concrete of the south-eastern abutment was placed directly on the smooth foliation surfaces of the schist. In the channel section, and the lower north-western abutment section, the concrete was placed across the foliation surfaces. The fault, known as the San Francisquito Fault, is inactive. It dipped 35–40° north-westwards, and consisted of up to 2 m of gouge. Fault movement had resulted in shearing and fracturing of the conglomerate. The conglomerate was a very low strength rock.

It seems (Ransome 1928) that no geological study was made of the site before design and construction, nor were any strength or durability tests carried out on the foundation rocks. The storage filled early in March 1928, and the dam failed shortly thereafter, the failure beginning on the north-western abutment. After the failure, only one small (channel) section of the dam remained standing (Fig. 9.1), and blocks of concrete of up to 10 000 t were swept up to 1 km downstream.

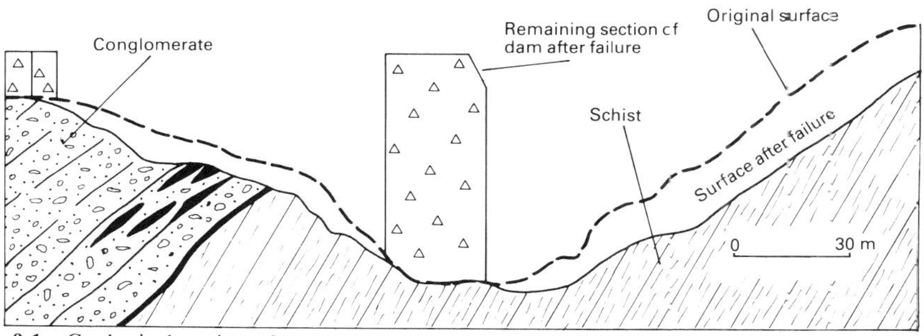

Fig. 9.1 Geological section of St Francis Dam site, California, showing the situation after failure. [After Ransome (1928).]

Up to 20 m thickness of foundation rock was scoured out in some sections. Subsequent study showed that the conglomerate was a compaction sediment, the matrix disintegrating when the rock was immersed in water, with a total loss of strength. Additionally, much of the matrix was soluble gypsum. Sliding, during the failure, also occurred along the foliation in the schist on the south-eastern abutment.

9.2.2 MEADOWBANK DAM, TASMANIA

The Meadowbank Dam is a massive buttress concrete dam, 43 m high, constructed on the Derwent River, Tasmania. The foundation rocks are Triassic sandstones, siltstones and mudstones, forming lenticular beds, dipping 3–4° north-eastwards, and with a downstream component of dip of 1°. Breccia zones occurred, which cut across beds, and carried strong water flows, while the bedding planes were often marked by softening. The rock mass was broken, also, by open near vertical joints.

The site was investigated by surface mapping, diamond drilling and 48 in. diameter calyx drilling, a seismic refraction survey, and several test trenches and test shafts. Detailed testing of the rock materials was carried out in the laboratory, and *in situ* shear tests were carried out in the calyx drill holes. *In situ* tests on bedding plane seams gave values for C of 7 kPa (1 p.s.i.) and for tan ϕ of 0.25. Laboratory tests gave values of $C = 21$ kPa (3 p.s.i.) and tan $\phi = 0.36$.

Detailed engineering analysis followed to determine a suitable factor of safety against failure by sliding. The analysis showed the need for a massive buttress dam, rather than the originally planned gravity, and pre-stressed cable anchors were provided to supply the balance of normal load (Fig. 9.2). The general comment by the engineers was that the experience gained confirmed the importance of thorough geological investigation and particularly of trial excavations at a damsite (Maddox *et al* 1976).

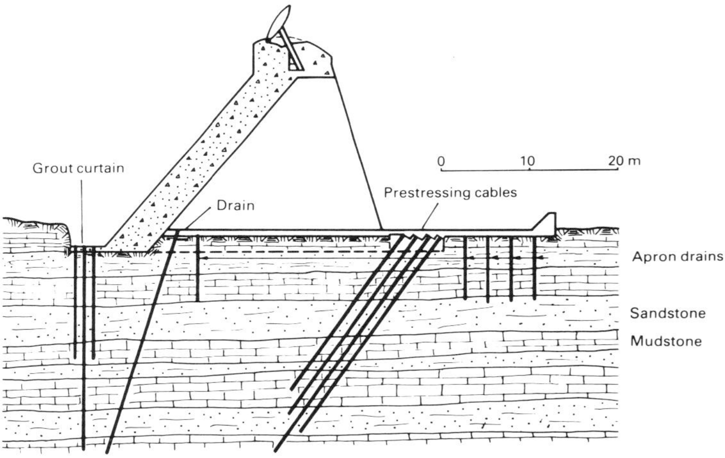

Fig. 9.2 Section of buttress foundation, Meadowbank Dam, Tasmania. [After Maddox *et al* (1976).]

9.2.3 MONAR DAM, SCOTLAND

Monar Dam was constructed on the River Farrar, Scotland. It is a double curvature arch dam, 161 m long overall and 35 m high. Initial investigations consisted of two shallow diamond drillholes and a surface geological survey. Subsequently, an exploratory tunnel was driven on each of the abutments, and further diamond drilling and trenching were carried out.

The foundation rock was a massive jointed granulite, largely unweathered, any weathered rock initially present having been removed in past glacial erosive activity. The granulite was intruded by a lamprophyric-type dyke, which it was thought might influence the stability of the foundations. Both *in situ* and laboratory tests were carried out to determine strength and elasticity. Tests to determine the dynamic elastic modulus gave field values of about 40% of the laboratory values, suggesting that jointing exerted a strong influence on the properties of the rock mass.

Four main sets of joints were recorded, one of which was almost certainly sheeting. These sheeting structures dipped towards the centre of the valley, were spaced at 30–45 cm and were smooth, although the open structures were filled with debris. Joints of the other three sets had dips of 40° or less, and one of these was also interpreted as a stress relief structure (Henkel *et al* 1962). The joints tended to be discontinuous. Many of the joint openings were clay filled. Friction tests on the joints gave values of about 50° for the angle of friction; clay filled joints gave average value of 25° and micaceous joints 30°.

Analyses of the effects of joints on the stability of the dam were carried out. It was assumed that, until movement occurred on the joints, the rock would behave as an elastic mass. A programme of grouting was carried out to provide a curtain to reduce seepage losses from beneath the dam; to fill the joints and so minimize deformation of the foundation; and to increase the shear resistance along the joints.

The stresses in an arch dam and its foundations are generally calculated by the designer on the assumption that the rock mass is elastic, homogeneous, isotropic, and continuous. It is, of course, necessary that the foundation stresses be less than the strength of the rock, but it is also necessary to ensure that no movement can occur on the discontinuities in the rock mass. It is critical in the geological investigation for an arch dam that the spacing, orientation and condition of all discontinuities be determined by the engineering geologist.

9.2.4 DARTMOUTH DAM, AUSTRALIA

The Dartmouth Dam is a multipurpose, earth and rockfill structure, recently constructed on the Mitta Mitta River, Victoria. The dam is 180 m high, 670 m long and contains 14.1×10^6 m^3 of material. The impounded water, of high purity, is to be used to boost water supply (via the Murray River) to South Australia; for irrigation; and for control of salinity in the Murray River.

The geological investigation involved mapping of the damsite at scale 1 : 1000, the drilling of 10 diamond drillholes (8 inclined, 2 vertical); the augering (using power augers) of 49 holes to refusal, 3240 m of trenching, and 13 seismic refraction traverses. Groundwater monitoring and a seismic risk appreciation constituted an essential part of the investigation.

The rock at the damsite is a closely jointed, sheared and locally foliated granitic rock, with an irregular pattern of weathering. Minor dyke intrusion had occurred. The geologists carefully mapped all structural discontinuities since it was considered that these would constitute the main foundation hazards insofar as seepage along them could result in erosion and piping. Permeable and compressible materials were removed and replaced with either concrete or well compacted fill. Where large areas of finely fissured rock occurred, they were covered with a layer of concrete or shotcrete 75 mm thick.

A number of major shear zones were present in the foundation and, some of which were normal, or nearly so, to the dam axis. In general, due to the high clay gouge content, these were impermeable, but the rock adjacent to them was closely jointed and highly permeable. Special attention was given to the grouting of these zones.

Sheeting structures were well developed on the abutments. These carried seams which consisted of weathered rock up to 2 m thick. These zones refused to accept cement grout, and had very low permeability. Nonetheless, concern was held regarding the erodability and compactability. Chemical grouting using a monomer AM9 effectively sealed this material as indicated by water pressure tests.

The problems of this very large earth dam centred on questions of permeability of the foundation, and in places, compactability. The site is deeply and irregularly weathered. The general foundation level was on a slightly weathered to moderately weathered rock mass, higher grades of weathered material being removed, generally, as well as specifically from structural defects where weathering of the rock material was high or complete (Fig. 9.3).

Fig. 9.3 Excavation for spillway, Dartmouth Dam, showing the highly weathered nature of the rock.

9.2.5 GENISSIAT DAM, AIN, FRANCE

Because of experience with the St Andalusia Dam in Spain, the Kentucky Dam in the USA, and a number of other damsites on limestone, engineers are reluctant to consider dam construction on, or close to, a limestone foundation.

For this reason, the Genissiat Dam at Ain, in France, is of interest. The dam is a concrete gravity type, 104 m high. The reservoir extends back from the dam for a distance of 23 km.

The dam is located on a synclinal fold in Cretaceous limestones, overlain by Tertiary lime rich sandstones, clay marl and thin limestones. The walls of the valley are covered by thick talus. As was anticipated, the limestones have a number of quite large solution cavities, but these were filled with what proved to be impermeable debris.

As part of the site investigation, a shaft was sunk to 40 m below the bed of the river, from which exploratory tunnels were driven to test mass permeability. It was found that the permeability was not high, and the initial concern was eliminated. 'Buried valleys' posed an additional problem, but it was found that the material filling these was also impermeable.

The alluvium in the floor of the valley required consolidation to withstand the load of the temporary diversion dams. Steel sheet piling was driven to 9 m, and clay-silicate grouting was carried out. Any cavities in the limestone were washed out and filled with concrete. The dam was successfully constructed, and in-service behaviour is satisfactory. This is considered to be due largely to the synclinal structure.

9.3 Buildings

Foundations for buildings are usually (but not invariably) placed at shallow depth, and are therefore on soil or rock in an advanced stage of weathering. Foundation pressures are relatively low. Where the foundation materials have relatively high strength and good deformation properties, the foundation type is a strip footing, which may by continuous, or isolated. Such foundations support one or several columns. If the properties of the foundation materials are inferior, it is the practice to adopt mat or raft foundations, which spread the load over the whole area of the building. In cases of very inferior foundation materials, piles are used: these may be either pre-cast, or cast in place.

Sometimes problems arise from factors other than the strength and deformation characteristics of the foundation materials. In cases in which the soils are expansive types, relative movement may occur due to wetting and drying, which sets up stresses in the structure, resulting in serious structural damage. Buildings located on unstable slopes are also liable to damage, or destruction, in the event of slope failure.

9.3.1 THE TRANSCONA GRAIN ELEVATOR, WINNIPEG, CANADA

The initial investigation of the site appears to have been restricted to plate bearing tests at the surface. It is now known that this is an unsafe practice unless the test results are interpreted in terms of the geology of the site. Beneath a loaded area, the load is spread, in a cone, with slope of about 1 : 1. If a small test plate is used, it is investigating only the characteristics of the material to very shallow depth, and, if inferior material occurs below this depth, which will be called on to carry the load of the much larger completed structure, the surface plate bearing tests will give no indication of its load bearing characteristics.

At the site of the elevators, a desiccated surface layer of glacial lacustrine deposits overlies very low strength saturated clays. Subsequent tests showed that the surface material had strength of 160 kPa, while the deeper material had strength of only 55 kPa. These weaker clays, it was found, would be influenced by the load of the large area raft foundation.

Foundation failure (October 1913) resulted in the tilting of the 6-bin elevator, which came to rest at an angle of about 27° from the vertical (Fig. 9.4b). The elevator suffered no structural damage and was later jacked to a vertical position, and supported by piers sunk down to hard rock.

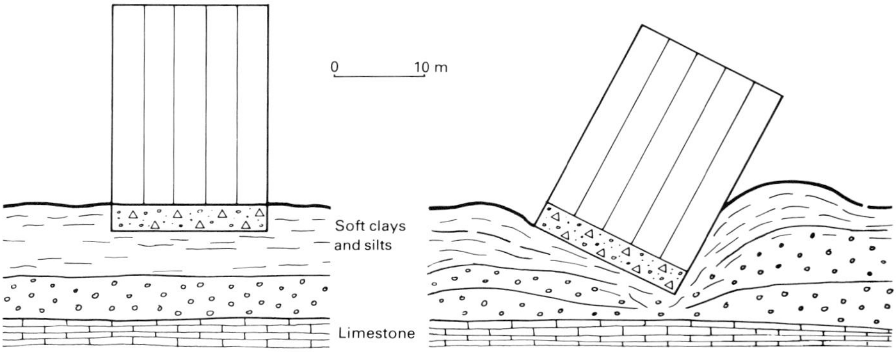

Fig. 9.4 Transcona Grain Elevator, Winnipeg, Canada, showing geological conditions and mode of failure.

9.3.2 YALLOURN POWER STATION, AUSTRALIA

This large brown-coal fired generating station is located in the Latrobe brown coalfields of Victoria. At the site, the brown coal and associated clays and sands have undergone monoclinal folding so that the beds are near vertical to vertical, and in places slightly overturned. The coal measures have been eroded and a thin overburden of sandy clay, and gravels has been deposited on the erosion surface.

Investigation of the site was carried out using percussion drilling, and some 40 holes were drilled. A contour map of the surface of the coal measures was prepared, and, after a study of the properties of the coal measures materials (brown coal, sand, tough ligneous clay and ligneous sand) and of the overburden, the decision was made to use the latter as foundation for about half the station, and, where the overburden was thin, to utilize the coal measures.

Because of the structural conditions, the foundation materials in the coal measures were highly variable, trending diagonally across the station site. Isolated footings were used to support columns, and such installations as boilers and turbogenerators; because of the variation in foundation materials, each footing required detailed consideration to avoid differential settlement (Beavis 1975). The foundation conditions for the two turbogenerators are shown in Fig. 9.5.

The main problem concerned foundations in the overburden where this material rested on brown coal. The weathered surface of the coal was in a thixotropic condition, and liquefied when subjected to vibration. For this reason, foundation levels for some of the footings had to be lowered from the overburden onto fresh coal.

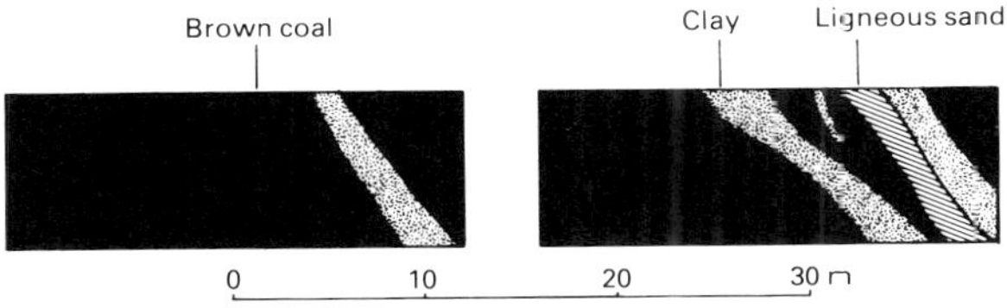

Fig. 9.5 Geological conditions in foundations for turbogenerator blocks, Yallourn D Power Station, Victoria, Australia.

9.4 Road and airfield runways

In terms of foundations, we are concerned in the case of roads and airfields with transient and impact, i.e. dynamic, rather than static loads. In the case of airfields, apron area foundations are also subjected to static loads. Geological factors are involved not only in foundation problems but also, in the case of roads, with problems of location, terrain stability, stability of constructed slopes and fills, and the supply of construction materials.

Location of highway routes requires a knowledge of the regional geology, combined with a sound terrain evaluation programme. Data from these services are used to fix alignments, define unstable areas, and provide a basis for the location of construction materials. It is essential, from the points of view of planning, design, and economics to know the construction conditions, the equipment required, and the respective lengths of cut and fill. These can all be determined from a proper terrain evaluation.

In certain segments of the proposed road, more detailed investigations may need to be carried out. In the case of unstable landforms, geological data are required for stability analysis and for the solution of foundation problems. For proposed deep cuttings in rock, determination of the geometry and condition of structural defects, as well as details of the weathering profile will be required. In some arid and semi-arid regions where the soils may be saline, areas of salinity will need to be defined, since saline soils are notorious for causing pavement failure. Again, in some cold, wet areas, swamp deposits and bogs present difficult foundation problems.

9.4.1 WINDY POINT CUTTING, OCEAN ROAD, LORNE, AUSTRALIA

Movement of a cut slope, and of the hillside above it, threatened to destroy a section of a main tourist road. The rocks were Cretaceous sandstones, dipping at 27° towards the resort beach. The beds were broken by vertical to near vertical joints spaced at from 1 to 10 m, and which showed progressive opening as the slow movement along a bedding plane progressed (Fig. 9.6). Detailed study of

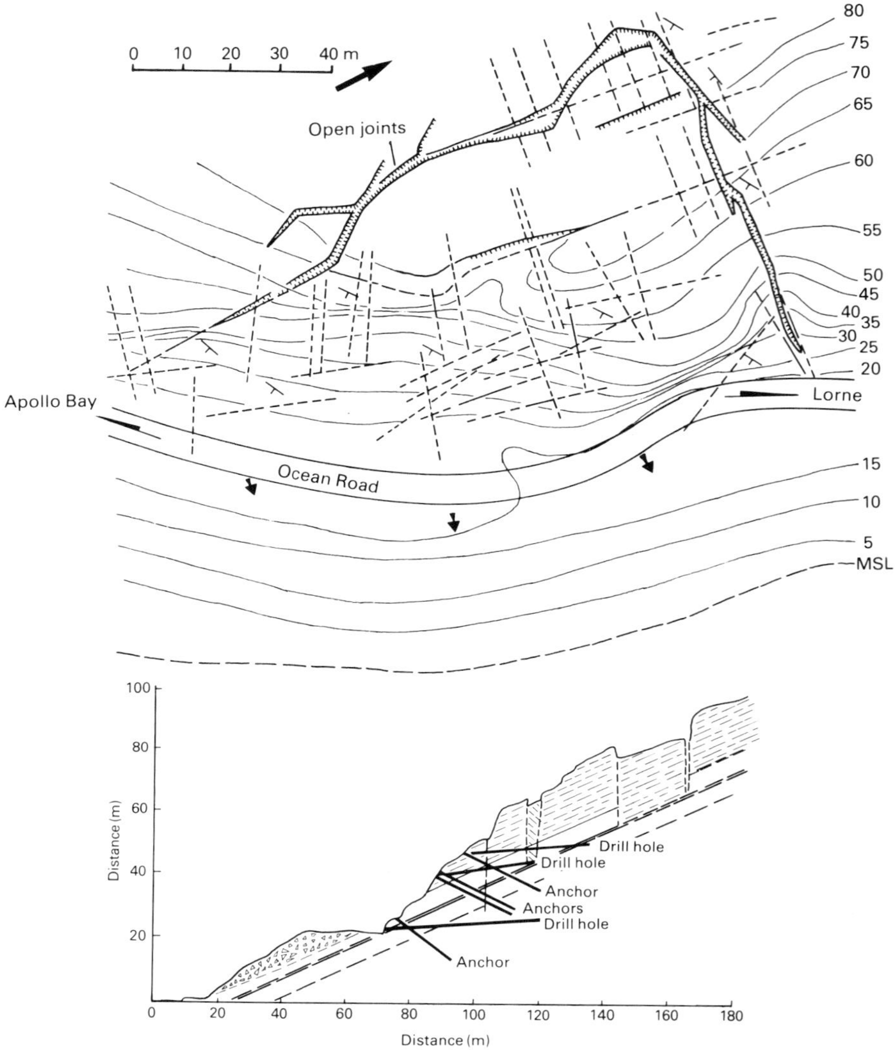

Fig. 9.6 Structural geology at the site of slope failure, Great Ocean Road, Lorne, Victoria, Australia.

movement patterns, and of joints showed that movement was occurring by slip along a bedding plane which outcropped at road level, and again, just below sea level.

Analysis of movements, and of the forces involved, resulted in the placing of 45 stressed rock anchors which tied the moving wedge of rock into the stable underlying rock mass. In addition, gently inclined boreholes were drilled to provide drainage. Monitoring of the cutting over 10 years has provided no evidence of further movement.

9.4.2 MOTORWAY ON BAGLAN MOORS, WALES

North of Port Talbot, Wales, the Port Talbot By-Pass Road crosses the Baglan Moors for a length of 2 km. The deposits are estuarine and comprise silts, soft clays and peat, with a thickness of some 15 m, resting on glacial deposits. Where buried tributary valleys join the main buried glacial valley, the superficial deposits are much thicker. Along the line of the road, the surface clays had reasonable strength, but peat beds up to 10 m thick had virtually no strength and were highly compressible.

Any embankment placed on the compressible peat beds would have resulted in unacceptable settlements. If displacement of the peat had been attempted by surcharge, an adjacent railway embankment would have been endangered. The solution adopted was to place a reinforced concrete raft, 36 cm thick and 263 m wide, supported by five rows of 45 cm diameter cast-in place piles, which rested on bedrock, at 2.3 m centres (Rogers 1972).

9.4.3 KINGSFORD SMITH AIRPORT, SYDNEY, AUSTRALIA

Reconstruction of the Kingsford Smith Airport, to accommodate modern aircraft, required a diversion of the Cook River, and extension of the main north–south runway into Botany Bay. The material at the site is a medium grained sand, locally overlain by silty, organic alluvium and overlying estuarine silts and clays.

The main interest in this project lay in the Botany Bay extension of the runway. Test boring showed that the materials had satisfactory strength and settlement characteristics, and so the embankment was placed directly on the sands. The initial material used for construction was dolerite, partly in a fresh condition, partly slightly weathered. In the early stages of construction, it was observed that the slightly weathered dolerite generally, and the fresh dolerite in part, was rapidly disintegrating. Research showed that a degraded chlorite, present in small proportions, but disseminated throughout the rock, expanded when in contact with sea water, and caused disintegration of the rock. The material was abandoned, and fresh basalt was used, although it was necessary to transport this about 100 km.

9.5 Bridges

Bridges are built as crossings over rivers, marine inlets, railways and roads. There are a number of designs, but the most important are the simple beam, cantilever, arch, rigid frame, and suspension. All types exert a vertical load on the foundation, although the load from the arch type is inclined, and can be resolved into the horizontal and vertical. The foundations of the rigid frame bridge exert both vertical and horizontal loads, while in a suspension bridge, the loaded cables tend to pull the supporting towers inwards, so that, to oppose this tendency, the towers are strengthened by cables anchored in rock.

Apart from geological considerations in the design of foundations, for bridges crossing rivers or marine inlets, attention has to be paid to the effects of currents and tidal movements on the foundations of the piers, and to the effects of narrowing the stream channel which may occur in some designs. In other words, consideration has to be given to the erodability of both bed and banks.

9.5.1 SYDNEY HARBOUR BRIDGE, AUSTRALIA

This bridge is an arch-span suspension type, construction of which was completed in 1932. It carries two rail tracks, a pedestrian walkway, and 6 lanes of motor traffic. Overall length is 1150 m. The rocks of the area are Triassic Hawkesbury Sandstone which is a unit comprising thick, horizontal beds of crossbedded sandstone of varying texture, with thin beds of soft, highly compressible shale. At the surface and close to the surface, the rocks are completely to highly weathered, and in this condition have strength of only about 40% of the fresh rock. The main geological factor, apart from weathering, was the possibility of sliding of foundations on, and inadequate strength of, the shale beds.

Initially, the foundations were to be placed 5 ft (1.5 m) into rock, with a maximum load of 31 tons/sq. ft (3.3 MPa). However, as a result of observations and tests during excavation, which showed the characteristics of the weathered rock to be unsatisfactory, excavations were made to 25 ft (7.6 m) into rock, and the foundation area increased to reduce the load to 13 tons/sq. ft (1.4 MPa). Each foundation was designed to carry a thrust of just over 20 000 t inclined at 45°, and a vertical load from the pylons of 11 000 t.

9.6 Underground openings

Tunnels are constructed for a variety of purposes: to carry traffic, water, power cables, and to provide access to large underground chambers such as hydroelectric power stations. Excavations may be in rock ('hard ground') or soils ('soft ground'), and the techniques of construction will depend on the type of ground. Geological factors play a significant role not only in the method of excavation, but in the shape of cross-section and on the stability of the excavation. While some tunnels are designed to be fully lined, others are designed to be unlined. In this latter case, safety, and efficient operation may require that lining be placed where the rock conditions are such that failure might occur.

It is often difficult to justify economically a detailed geological investigation for a tunnel, unless the tunnel is to be at a shallow depth. For deep tunnels, drilling investigations are certainly uneconomical, and most investigations in such a case consist of a surface geological survey supplemented by geophysical investigations and a limited amount of drilling at portal sites, or at points where some critical geological problem may require solution. On the other hand, for large openings, such as underground power stations, very detailed geological investigations are essential.

Probably the three most important geological factors to be considered in hard rock underground excavations are rock weathering, in so far as quality of rock, and sound cover are concerned; structure, which will control the stability of the excavation; and groundwater hydrology. The nature, condition, and (in relation to the orientation of the excavation) the attitude of structural defects are critical factors in determining the shape and stability of the excavation. Experience has shown, for example, that joints which are parallel to the long axis of an excavation can create instability of walls.

Groundwater assumes considerable importance in so far as it, too, can influence stability. Inflows of any magnitude can cause hazardous working conditions, and, in a power station, damage to machinery. The development of groundwater pressure behind tunnels and power stations can also result in failure of the lining. This is especially the case for steel lined pressure tunnels.

In soft ground, the problems are entirely different from those in hard ground tunnels. All soft ground tunnels are lined. The major problem is the controlled excavation of some materials, such as clean dry (running) sand, and saturated flowing sand. Under these circumstances, chemical grouting, or forepoling will have to be used, or the tunnel may have to be driven by shield and compressed air.

9.6.1 COOLING WATER TUNNEL, YALLOURN D STATION, AUSTRALIA

A tunnel to serve as an inlet for cooling water was constructed in both hard ground and soft ground at Yallourn. Brown coal, sands and clays of lower Tertiary age overlie, disconformably, Cretaceous sandstones and shales. Both units are monoclinally folded, with dips of from 20° to 90° in the Cretaceous beds, and from 70° to 90° (and overturned) in the coal measures. The brown coal and Cretaceous sediments were strongly joined with fracture spacing index 10 cm.

Because of the shallow depth (18 m) and the location of the tunnel in relation to the foundations of existing stations, a detailed geological investigation was completed along the full line of the tunnel. The upstream section of the tunnel was excavated in Cretaceous sandstones. Pneumatic shovels were used since the highly jointed nature of the rock did not require the use of explosives. The condition of the rock, however, required heavy steel ribs for support, some of which failed by buckling. Within the coal measures, the sections in brown coal and clay stood for at least two weeks without support; concrete lining was placed at two weekly intervals. Saturated sands were in a flowing condition, and sections of the tunnel in this material were driven using bulkheads, and forepoling, with the advance being by heading and bench methods. The groundwater in the sands contained H_2S in solution which was released in toxic quantities, so that additional forced ventilation was required.

9.6.2 QUEEN'S – MIDTOWN TUNNEL, NEW YORK

Driven under East River as a traffic tunnel, this structure is about 10 m in diameter and was excavated in both hard ground and soft ground, using a variety of methods including conventional drilling and blasting, shield, and compressed air techniques. Because several other tunnels existed in the area, the geology was known to some degree. However, diamond drilling was carried out to determine the rock level beneath superficial deposits. This work showed that the tunnel would be driven partly in rock (schist, gneiss dolomite and limestone) and partly in superficial soils, with the schist highly to completely weathered in some sections.

Excavation of rock utilized conventional tunnelling methods, but care had to be exercised in blasting to avoid damage to a nearby power plant. Methods involving shields and compressed air were used to drive through the glacial sands, clays and gravels. In one section, the tunnel encountered loose rip rap from older surface works. This had to be consolidated by grouting. Even so, this material (as did the sands and gravels) required support for safety.

9.6.3 POATINA UNDERGROUND POWER STATION, TASMANIA

This station, 99 m long, 14.5 m wide and 28 m high, is located 165 m underground in horizontally bedded, fine grained, dense massive mudstones, with thin beds of shale. The site is adjacent to a steep escarpment some 1000 m high. This is believed to have been responsible for the anomalous horizontal 17.7 MPa *in situ* stresses normal to the station axis, as against 8.6 MPa vertical stress and 12.5 MPa horizontal stress parallel to the station axis.

Tests for the strength of the rock material gave values of 34–69 MPa, and for the *in situ* strength 34–42 MPa. From analysis, the designers concluded that, in the existing stress field, a circular arch roof would be unstable. A flat roof would be stable, except that near the walls, stresses in the roof would exceed the strength of the rock. For this reason, a trapezoidal cross-section was adopted for the roof (Fig. 9.7).

Construction proceeded without difficulty. Support was provided by rock bolts, mesh and gunite. Although the design assumed that the rock was elastic, provision was made for creep in the machine foundations, which were mounted on sliding joints. A total movement of 4 mm occurred in the first year. No movement has been recorded since.

9.6.4 ROUNA 2 STATION, NEW GUINEA

The machine hall, located in agglomerate, 160 m below the surface, is 39.5 m long, 13.5 m wide and 21 m high. The geological investigation involved surface surveys and a total of 1080 m of diamond drilling. The agglomerate is coarsely stratified with layers of coarse angular agglomerate, with a matrix of welded tuff, and rounded agglomerate, with a slightly to moderately indurated tuff matrix. Lenses of tuff also occur, which are up to 5 mm thick. The beds occupy a gentle synclinal fold; with dips of from 5 to 10°.

Joints, which are open at the surface, tighten with depth. They consist of two steeply dipping intersecting sets, with the long axis of the station bisecting the intersection. Joint pattern and spacing varies from bed to bed. During construction, the joints tended to open, accompanied by minor crumpling of the rock adjacent to the joints. Although no groundwater problems occurred during construction, since the station was placed in service, seepage at the rate of 70000 l h^{-1} is occurring from the tuffs and more permeable agglomerates. The roof is stabilized with rock bolts, and bolts were used to strengthen the walls. The

Fig. 9.7 Roof of Poatina Underground Power Station, Tasmania.

whole roof and the upper parts of the walls are protected by mesh and gunite. The lower parts of the walls are inclined.

9.7 Coasts and harbours

Two basic geological problems are involved in beach and harbour engineering: stability of bays, beaches and cliffs; and foundations for engineering structures. The former problems centre on geological processes such as erosion, transport and deposition; the latter on the nature and properties of the sea floor. Long experience with harbour works has shown that unless natural sediment movement is considered and the effect of harbour works on this movement taken into consideration in design, the harbour is almost certain to silt up rapidly, and to require frequent dredging to maintain its utility.

9.7.1 PORTLAND HARBOUR, AUSTRALIA

In an earlier chapter, the erosion of headlands and the transport of the eroded material by longshore currents was discussed. As part of the preliminary geological investigations for the new Portland Harbour, tests were carried out using marker minerals (mainly garnet) to determine the direction and rate of sediment movement (Fig. 9.8). The conclusions drawn from this work were that

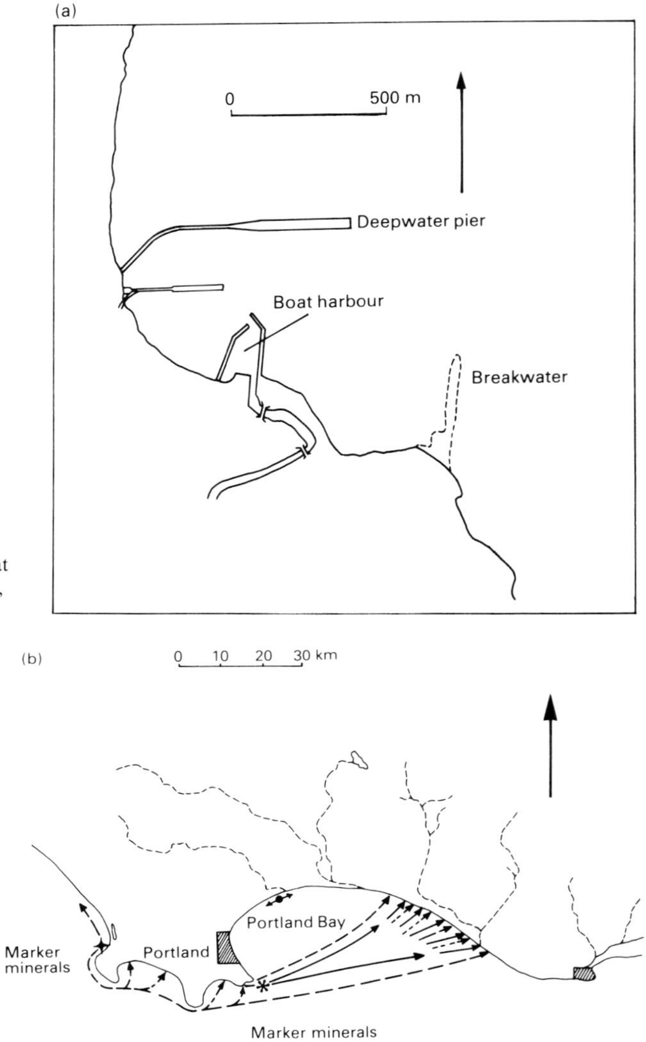

Fig. 9.8 Sand movement at Portland Harbour, Victoria, Australia.

the sand in the proposed harbour area had an almost entirely local origin and that only a minor amount of sand would enter the harbour due to long shore drift. In the rough winter season, in particular, virtually all transported sand bypasses the harbour area. It was found that a seasonal interchange occurs between harbour floor and beach, so that only a very minor amount of dredging is required in the harbour area. The design of the deepwater pier was modified to prevent the accumulation of sand.

9.7.2 THE KING'S WHARF, SUVA, FIJI

The site of the wharf, which replaced one destroyed in an earthquake, consists of poorly consolidated marine sediments overlying a very stiff to brittle mudstone. Some fill from the earlier harbour works overlies the marine sediments. The wharf is supported on hollow concrete piles driven through the fill and the marine sediments into the mudstone bedrock. The piles averaged a length of 40 m.

A curtain wall at the back of the wharf was constructed using steel sheet piling which, for stability, required the placing of backfill. The surcharge imposed by this backfill caused some concern for the safety of the foundations of proposed buildings nearby. For this reason, these were changed from shallow strip footings to pile and beam support.

9.8 Engineering materials

The search for construction materials, whether soil or rock, is based on a knowledge of the regional geology, and weathering patterns. Irrespective of the purpose for which the rock material is required, any quarry site will be located in areas of shallow weathering, while the purpose for which the rock is required will often determine the type of rock selected for development. Soil borrow areas will be located either in areas of deep rock weathering, or in areas of extensive deposits of transported soils. Whether it is rock or soil which is being sought, final site selection will be made only after geological surveys supplemented by subsurface investigation to prove the extent of the deposit, and laboratory testing to prove the suitability of the materials, and to determine their physical properties.

9.8.1 ROCK MATERIALS

Rock which is to be used as concrete aggregate should meet certain physical and chemical standards. It must be fresh, have high strength and durability, and produce a crushed product which does not contain a significant proportion of platy or elongated particles. It is desirable that the particles have a reasonably particle shape when the rock is broken or crushed: for example, flat, slabby or platy fragments are almost invariably unsatisfactory, irrespective of the purpose for which the rock is required. These requirements are more likely, but not necessarily, to be met in igneous rocks.

Rock which is to be used as concrete aggregate should meet certain physical and chemical requirements. It must be fresh, have high strength and durability, and produce a crushed product which does not contain a significant proportion of platy or elongated particles. It is desirable that the particles have a reasonably rough surface, to achieve an effective bond with the cement mortar; for this reason, some river gravels are undesirable, due to the very smooth surface of the pebbles. In chemical terms, the rock should be free from materials which may react with the cement to produce a deterioration in the concrete.

It is known that some materials are deleterious in concrete, e.g. organic materials and sulphates. It is also known that some minerals react with some types of cement to cause expansion and disruption of concrete: this is known as the alkali–aggregate reaction. If the aggregate contains acid volcanic glass, opal, chalcedony or other cryptocrystalline forms of silica, and the cement has an alkali (Na_2O + soda equivalent K_2O) content of more than 0.6%, this reaction, *may*, but does not necessarily occur. It is essential, therefore, that all aggregates, including sands, be examined carefully to determine whether or not these potentially harmful minerals are present. If they are, then the aggregate should be rejected, or the engineer must ensure that a low alkali cement is used.

Road aggregates have requirements which vary with the position they will occupy on the pavement. Load bearing aggregates carry the traffic loads and transmit them to the subgrade. These require high strength and high abrasive resistance, and to have an angular particle shape. The basic igneous rocks have proved to be the most satisfactory. Aggregates for bituminous mixes for the surface course require to be resistant to abrasion, and to have qualities which ensure a permanent bonding with bitumen. It is desirable, also, for safety reasons that these aggregates be dark in colour, and to have angular particle shape. Again, the basic igneous rocks, especially basalt, have proved to be the most satisfactory.

Rip rap is used as a protective layer of rock on embankments in order to protect the soil of the embankment from erosion, and to increase stability. It is normally required that the broken rock have a shape which produces a stable interlocking of the boulders, while the rock material must be durable under the conditions to which it will be exposed. In particular, resistance to alternate wetting and drying is frequently a prime requirement.

9.8.2 SOILS

Soils used for engineering purposes may be either residual soils, produced *in situ* by the weathering of rock; or they may be transported soils which have been transported to their present location by wind, water or ice. The nature, composition and homogeneity of residual soils will depend on these qualities of the parent rock mass, and on climatic factors. Transported soils are usually highly variable in both texture and composition in any one deposit.

Essential to the investigation of a potential soil borrow area are the definition of the lateral and vertical extent of the soils, and the delineation of soil types based on texture, composition and relevant engineering properties. When a soil is disturbed and reworked, a new fabric is imposed, which has a significant effect on its engineering properties. This is a factor which has to be borne in mind when carrying out tests on soils.

References and further reading

Anderson J. G. C. & Trigg C. F. (1976) *Case Histories in Engineering Geology*. Elek Science, London.

Anon. (1980) Dartmouth Dam: geological report. State Rivers and Water Supply Commission, Victoria.

Beavis F. C. (1975) Engineering geology of 'D' power station, Yallourn, Victoria. *Quart. Jl. Engng Geol.* **8**, 103–17.

Engineering Geology Specialist Group (1980) Papers, Series 1. Geological Society of Australia, Sydney.

Engineering Geology Specialist Group (1982) Papers, Series 2. Geological Society of Australia, Sydney.
Henkel D. J., Knill J. L., Lloyd D. J. & Skempton A. W. (1962) Stability of foundations of Monar Dam, pp. 425–41. Trans. 8th Congr. Large Dams, Edinburgh.
Institution of Engineers, Australia (1969) *Engineering Geology*.
Jaeger C. (1979) *Rock Mechanics and Engineering*. Cambridge University Press, Cambridge.
Maddox J. M., Kinstler F. L. & Mather R. P. (1976) Meadowbank Dam: Foundations. Civ. Eng. Trans. Inst. Engrs. Aust. pp. 321–333.
Ransome F. L. (1928) Geology of St Francis dam site. *Econ. Geol.* **23**, 553.
Rogers S. H. (1972) Foundation problems of motorway construction. *Quart. Jl. Engng. Geol.* **5**, 145-58.

10

Environmental Geology

10.1 Introductory remarks

While it is true that geologists have always been concerned with the effects of mining and civil engineering works on the general environment, it is only in recent years that systematic study of environmental geology has been undertaken. Significant advances have been made in the application of geology to regional and urban planning, to the disposal of wastes of all kinds: solid, liquid, and radioactive, and to the general effect of engineering works such as large dams. Environmental issues often tend to be highly emotive; it is essential, therefore, for the engineer and geologist to approach these issues as objectively as they approach other problems in their professions. The solution is often compromise, based on a balance of scientific, engineering, social, economic and political considerations.

Environmental problems are clearly due to a number of factors, the more important of which are population growth and advancing technology, together with the affluence which the latter has tended to produce. The problems are manifold; but the geologist is primarily concerned with those associated with 'solid' earth and groundwater pollution. In studying these problems, almost every branch of geology plays some part: mineralogy, petrology, structual geology, stratigraphy and geomorphology, as well as applied aspects such as hydrogeology.

10.2 Geology and regional planning

Haphazard development of regions for residential and industrial purposes is no longer socially acceptable, since, without careful investigation and planning, serious environmental problems, which can be avoided, might arise in the future. The basic geological investigation has as its aim the recognition of areas of unstable landforms, varying foundation conditions, potential flooding and other natural hazards, exploitable soil and rock, in addition to groundwater studies and the siting of required engineering works such as water storages, sewerage treatment plants, and waste disposal, both domestic and industrial. With basic data provided by the engineering geologist, it is possible for planners and engineers to define potential residential and industrial areas, and to locate those environmental damaging activities, e.g. quarries, sewage treatment works, and waste disposal sites, in areas where least harm will be done to social amenities. In carrying out the necessary surveys, the geologist must be aware of both the general issues, as well as specific issues. For this reason, the surveys will be, in the first instance, general and regional. The more general investigations will then be followed by detailed study of the specific problems.

A general, regional survey for environmental geological purposes defines areas of rock outcrop and of superficial deposits, and, for the former, describes rock types, outcrop and weathering patterns, age, relationships and structure. Topographic expression of the various rock units, the soil types developed on them, and characteristics of the soils are also described. This work permits the recognition of environmental units (Fig. 10.1, Table 10.1). The example

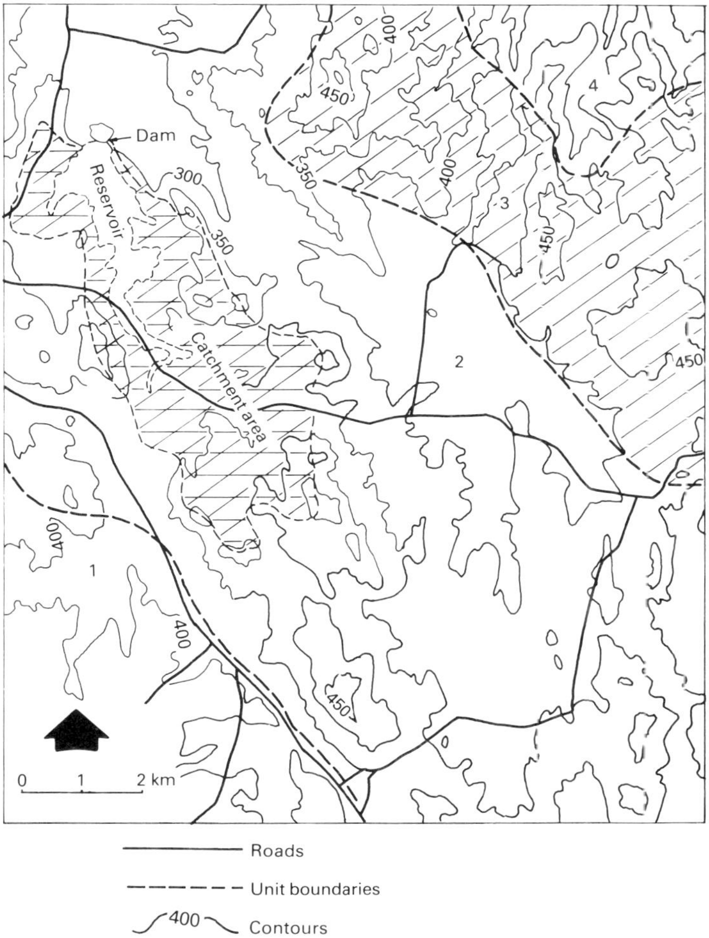

Fig. 10.1 Environmental units as defined in Table 10.1 for an area in the Bathurst Orange Growth Centre, New South Wales, Australia.

illustrated was part of the survey for a regional growth centre, and lies on the outskirts of the area studied. Interest was centred in the development of market

Table 10.1 Diagnostic characteristics of the four environmental units of Fig. 10.1.

E.R.U.	Topography	Geology	Soil types	Vegetation	Engineering problems
1	Gently undulating, relatively low lying area. Poorly dissected.	Minor scattered outcrops of basalt.	Red earths and black alluvium	Mainly improved pastures. Intense agriculture. Yellow box timber on red earths.	Erosion is negligible. Some problems with montmorillonite — a product of basalt.
2	Andesites forming ridges. Meta-sediments forming valleys. Acid intrusives forming hills.	Meta-andesites, tuff, breccias, and wide variety of meta-sediments. Ultra basics and acid intrusions.	Predominantly red earths, with black alluvium along creeks.	Varies considerably. Timber is mainly peppermint and brittlegum.	Only moderate erosion. Sulphides very common on fractures in andesite — not suitable for concrete aggregate.
3	Heavily dissected region of deep gorges and steep ridges.	Foliated and meta-morphosed shales, siltstones and mudstones	Soils generally very thin. Mostly red in colour where developed.	Timber is mainly wattle, brittlegum and stringybark.	Meta-sediments highly cleaved. Erosion only slight.
4	Acid tuffs give gentle, undulating country. Slates give prominent ridges.	Acid tuff and flow-dacites, quartz-feldspar porphyries. Minor slaty units.	Predominantly yellow solodic soils. Colluvium is common on lower slopes.	Yellow box on water courses. Red gum on slopes. Stringybark on ridge tops.	Tunnelling in solodic soils Constructional materials are limited.

gardens and extractive industries in this segment of the proposed growth centre.

When planners have laid out a broad design for development, more detailed environmental geological studies will be required. These will include detailed geological, geomorphic and hydrogeological mapping and geotechnical investigations. Specific area maps and reports will be prepared as a basis for detailed design and construction of the various elements of the project.

10.3 Urban planning

The aim of urban planning is to produce a design of residential areas, work places, recreation areas, transport and other service facilities so as to utilize land resources in the most beneficial way in both the short and long terms. One of the most important aspects to be considered is the effect of changing the patterns of land use.

With urbanization, a number of critical changes are induced in the environment, including hydrological characteristics, which, in turn affect other elements. As land use changes, run-off behaviour will change: e.g. natural grassland and forest catchments have a much lower run-off coefficient, and higher infiltration factor than areas which are paved and covered by buildings. This has two important effects: a reduction in groundwater recharge, and an increase in flood potential. For this reason, natural drainage lines in small catchments should be left in their natural state. Further effects are the dispersion of bared soils, during development, by raindrop impact, and a sharp increase in sediment yield from drainage basins.

It is imperative, for urban planning, that the suitability of the area be assessed in terms of stability. Areas of actual or potential slope instability are to be avoided. Lives and property are at risk if homes are built on unstable slopes. The actual preliminary construction of roads and services may aggravate the situation, while after construction, increase in water intake to the slope from house drainage and septic tanks may further aggravate the position. If a slope is potentially unstable, careful attention must be given to domestic drainage. In the Camden urban development on the outskirts of Sydney, one housing area was located on slopes with a long record of failure. One new road had to be abandoned, and several homes are endangered. The disastrous landslide in Dunedin, New Zealand, in 1979 (Fig. 10.2), illustrates the danger of using unstable slopes for urban development.

Two primary factors concerning environmental damage, resulting from urban development, have to be considered. The first of these is pollution of surface waters and groundwater. Biological contamination is possible from septic tanks, while chemical pollution from detergents, and industrial wastes will almost certainly occur. In addition, recent data suggest that contamination of soil and water by lead from car emissions is becoming a problem in urban areas.

The second factor which has to be considered is what is called the sterilization of land resources. Land utilized for urban development immediately becomes unavailable for any other purpose — extractive industry, farming, or forestry.

Unstable areas due to unnatural causes also have to be considered. For the

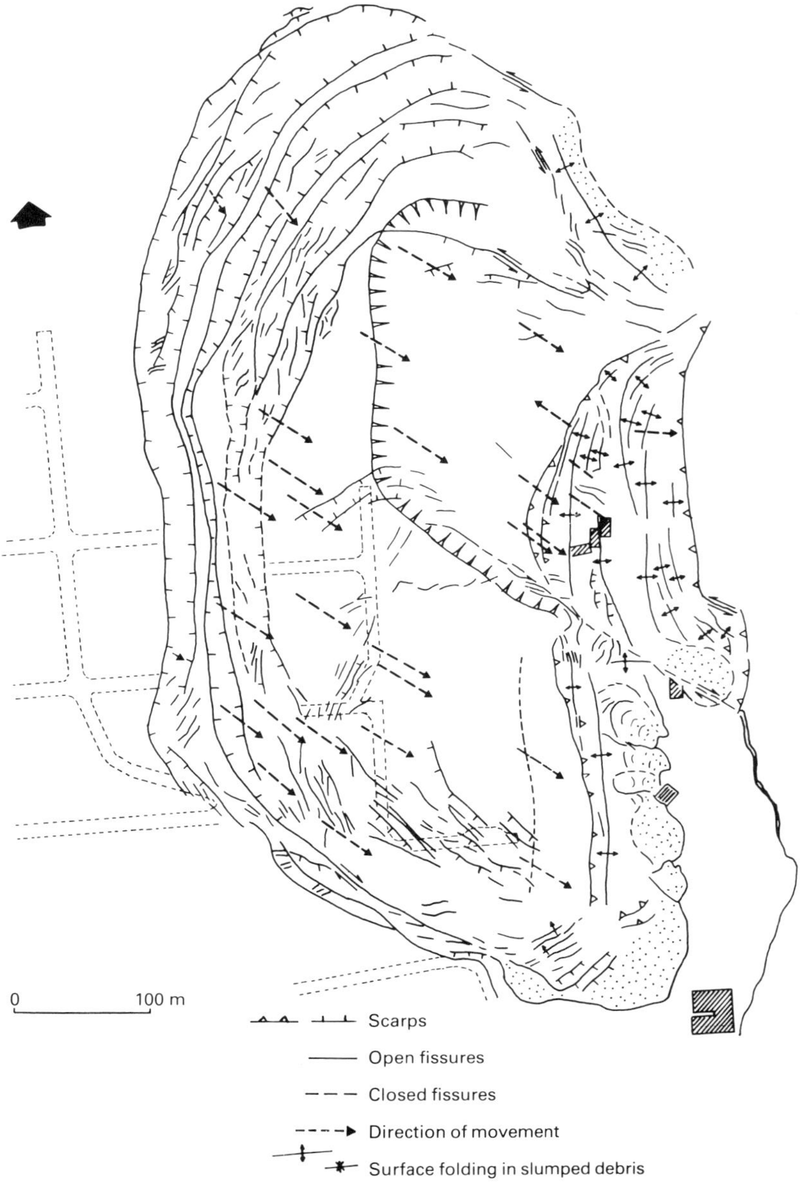

Fig. 10.2 Map of the disastrous landslide of 1979, Dunedin, New Zealand. [After Coombs (1981).]

most part, these are areas of subsidence due in some rare cases to the excessive removal of groundwater, or, more commonly, to mining activity. Some areas in California have suffered severe subsidence due to oil extraction, but subsidence is almost universal in coal mining areas. It is not generally appreciated that ground movements are also associated with open-cut coal mines. In Germany and

Victoria, quite strong horizontal, and lesser vertical, movements have been recorded in a zone up to 100 m from the margin of open cuts. These movements result in damage to water supply and sewerage services, and buildings.

One important geological factor which has to be considered in urban development is the suitability of the soil and rock for foundations. Information is required on the strength, deformation, and water-reactive characteristics of soils and rocks; the thickness of soils, and the weathering condition of the rocks. Additionally, information on groundwater conditions will be necessary for the design of deep foundations. Areas with specific foundation problems (e.g. highly compressive soils, swelling soils) must be clearly defined.

Finally, in urban planning, geological factors will influence location of services: waste disposal, water supply, transport corridors and roads, and construction materials. Waste disposal involves both domestic and industrial; and solid and liquid wastes. Solid wastes can be used to fill old excavations such as quarries and sand pits, and, if the wastes are properly placed and compacted, it is possible to put these 'landfill' areas to use at a subsequent date. It is important, however, to ensure that pollution of streams and groundwater does not occur. In the Menai waste disposal area, south of Sydney, Australia, geological studies have been conducted to ensure that pollution of the Georges River, with its extensive oyster farms, does not occur (Knight *et al* 1978). Liquid waste (sewage) disposal may be by reticulated sewerage, or septic tank, system. For the former, careful location of treatment works is required, as well as knowledge of the geology of the urban area. This latter will control, to a high degree, the costs of the system. For the septic tank system, the main considerations are the permeability of the soil, and the likelihood of biological pollution of groundwater. Industrial liquid wastes are commonly disposed of by well injection methods. In the past, unplanned well injection of these wastes has effectively destroyed urban groundwater basins.

Water supply services can be of two main kinds: large surface storages, often with necessary purification installations, or groundwater. Large surface storages are costly, and tie up quite large areas of valuable land. Moreover, the supply is frequently subject to climatic variations, with shortages in times of drought. On the other hand, groundwater supplies require a much lower capital outlay, and provide a regular supply. In this case, however, the resource is only very slowly renewable. Economic factors are probably decisive. For example, in parts of New Guinea, it was found to be more economical to use groundwater, increasing the number of bores as the urban area developed. The same procedure is adopted in a number of developing countries, such as Indonesia.

The provision of transport facilities involves geology both directly and indirectly. Directly, geological data are required for road location, including the foundation conditions, and for the assessment of batter stability conditions with respect to erosion and slumping. Availability of construction materials also has to be taken into consideration. Indirectly, road location has to be considered in sociological and aesthetic terms: the possible destruction of significant sections of recreation areas, and of natural environmental importance. In this sphere, the geologist can give valuable policy advice.

10.4 Waste disposal

Brief reference has been made to the disposal of domestic and industrial wastes. Here, these problems will be considered in more detail. As well, the geological aspects of the disposal of radioactive wastes, a critical problem, will be discussed. With increasing urbanization, industrialization, and the continued use of radioactive materials, safe and environmentally acceptable methods of waste disposal have become important political and social issues.

10.4.1 DOMESTIC WASTE

Two main types of domestic waste have to be considered: garbage and sewage. In developed countries, the average garbage amounts to about 3 kg per person per day. Although some municipalities dispose of much of the garbage by burning, the most common practice is to place the garbage, with or without treatment, into the ground. This immediately creates a problem of pollution of surface and groundwaters, both chemical and biological. In California, it has been found that continuous leaching over a period of one year of so-called sanitary landfill (garbage covered with soil) will extract 1.5 t of Na^+ and K^+, 1 t of Ca^{2+} and Mg^{2+}, 0.9 t of Cl^-, 0.2 t of SO_4^{2-} and 3.9 t of HCO^-_3 per acre foot. Much of this passes into the groundwater body, rendering it unfit for consumption.

Because of these hazards, it is essential to ensure, as far as possible, that the base of the fill area is sealed to prevent pollution of aquifers. An associated geological requirement is suitable earth-fill material to serve as a cover (Fig. 10.3).

Fig. 10.3 Landfill disposal of urban wastes, Kainantu, New Guinea. (Photograph by courtesy of M. J. Knight.)

Sewage disposal ranges in type from the primitive pit latrine, through the septic tank, to chemical treatment, with settling tanks or lagoons of the high quality reticulated sewerage system. In every case, some pollution of ground and surface waters may result. The composition of domestic sewage, based on a water carriage capacity of 360 l / day, is shown in Table 10.2.

The potential of groundwater contamination can be assessed using geological and hydrogeological data (Le Grand 1964). Factors which have to be taken into account are depth to water table, sorption capacity of the soil, which

Table 10.2 Composition of domestic sewage.

Solids	mg/l total	organic	inorganic	5-day 20°C Biochemical oxygen demand (mg/l)	No. coliform bacteria per 100 ml
Total	800	450	350	200	10^{11}
Suspended	275	175	100	150	
Settleable	175	50	125	70	5×10^{10}
Non-settleable	100	70	30	80	5×10^{9}
Dissolved	525	275	250	50	

is a function of particle size, permeability of the soil, gradient of the water table, and the distance between the pollution site and the point of groundwater withdrawal. Ratings can be established which permit the fixing of areas of safe groundwater use.

10.4.2 INDUSTRIAL WASTES

Industrial wastes comprise three main types: solid, liquid and gaseous. The gaseous wastes are normally disposed of into the atmosphere, with consequent fall-out onto the ground, via 'acid' rain, producing acid soils. Solid and liquid wastes are placed on, or in the ground, using natural or artificially created space. Disposal on the surface inevitably leads to contamination of surface and groundwaters.

Surface disposal of wastes can produce readily observable pollution (Figs 10.4, 10.5, 10.6), but if the principles adopted for sanitary landfill methods are applied, this can be minimized.

Three main methods exist for the subsurface disposal of wastes:

(1) pressure injection of liquid wastes into *deep* permeable formations;

(2) incorporation of liquid wastes into cement slurry which is then injected into fractured, impermeable formations;

(3) storage in caverns in salt formations, e.g. in oil fields. This applies to solid wastes.

In all of these methods, knowledge is essential not only of such geological properties as porosity, fracturing, and permeability, but also of the likelihood of chemical reaction between the rock and the waste. Even then, other factors have to be taken into account. The Rocky Mountain Arsenal has produced chemicals for use by the United States Army. Initially, the contaminated waste water was disposed of in shallow wells, which caused pollution of the shallow groundwater. As a result, the decision was made to use a well 3675 m deep. The response to this was a series of earthquakes reaching a maximum intensity of 4.3 on the Richter scale.

10.4.3 RADIOACTIVE WASTES

Radioactive wastes are produced in two stages: at the mining stage, and as end products of nuclear reactor and related processes. At the mining stage, the mine

Fig. 10.4 Disposal of dairy wastes, Rochester, Victoria, Australia.

Fig. 10.5 Copper-chrome arsenate treated timber waste disposal, Colac, Victoria, Australia. (Photograph by courtesy of EPA, Victoria.)

Fig. 10.6 Uranium wastes, Rum Jungle, Northern Territory, Australia. (Photograph by courtesy of M. J. Knight.)

wastes, including tailings from treatment plants, are dumped on the surface and held in tailings dams. At Rum Jungle, Northern Territory, Australia, tailings are placed in a landfill dam. In the 1974–5 wet season, the dam overflowed, with the result that plants and fish were killed. At Grand Junction, Colorado, USA 300 000 t of tailings were used as landfill on which buildings, including a school, were constructed. Increased radioactivity was reported in nearby rivers, while reports also suggest an increase in infant death rate and an increase in pulmonary carcinoma. Buildings were removed from the landfill site.

Reactor and related wastes are defined in terms of levels of radioactivity, and dilution factors are prescribed for each before disposal (Table 10.3).

Table 10.3 Radioactive dilution factors.

Waste	Dilution factor	Source
Low level	1000	Nuclear reactor process water
Intermediate level	1000–100 000	Radio-isotope preparation laboratory
High level	>100 000	Fuel element and processing plant

Low level wastes are treated by flocculation and ion exchange, and are then discharged into drainage systems. This is unsatisfactory, since natural processes can reconcentrate the radioactive materials, e.g. by absorption onto clay minerals. Attempts have been made to dispose of intermediate level wastes by injection via wells, into 'safe' rock types. Other methods include injection in a cement slurry. At present, high level wastes are solidified in concrete, or adsorbed onto vermiculite or zeolites, stored in drums and buried underground. After 10 years most radioactive elements present have decayed to a low level. This does not take account of those elements with a long half-life.

Experimental work at the Australian National University has shown the possibility of a new method of disposal. A synthetic rock 'synrock' made of barium aluminum silicate, with oxides of zirconium, calcium, titanium, potassium and aluminium is highly resistant to leaching. Incorporation of wastes in synroc, which is then buried, insures against leaching out of radioactive materials and consequent contamination of groundwater.

10.5 Geological hazards

Geological hazards are a particular group of geological processes that endanger man and his structures. The main types are floods, rock avalanches, earthquakes, landslides, soil erosion, moving sand in arid regions, tsunamis and volcanic eruptions. The role of the geologist in the study of these environmental issues lies in prediction and control. In one sense disasters resulting from natural events could be said to be the fault of man himself: he is aware of the risk of floods, of earthquakes and volcanic activity, but he continues to develop areas subject to these disasters since social, political and economic pressures tend to prevail. Future developments should be determined on the basis of hazard assessment, to avoid disasters with their high toll of human life and of property.

Geological hazards are, to a large degree, incapable of prevention by human agency. In some instances, such as floods and soil erosion, management control after the first event may prevent a repetition. Hazard events are, for the most part, probabilistic in their frequency distribution, but almost invariably a high degree of uncertainty exists as to the specific timing of a future event. The event is usually one of sudden occurrence, and is usually spatially concentrated. Basically, the problem facing the environmental geologist is the recognition of the hazard, and the prediction of an event in space and time.

A number of approaches can be adopted. The most obvious is, having recognized the hazard, to avoid development in the area. A more usual approach is to accept that the hazard event will take place, and to design structures apporpriately, e.g. in seismically active areas. The third approach is to accept that the hazard event will occur, and accept the risk involved. Whichever is adopted, certain basic data are required as a basis for the decision. The data will include a statement of the hazard, the frequency of the hazard event, and the magnitudes of past events. These data can be recorded on maps (e.g. Fig 10.7). Prediction of future events, and their magnitude can be based on physico-mathematical models using the basic data, or on statistical evaluation of past events.

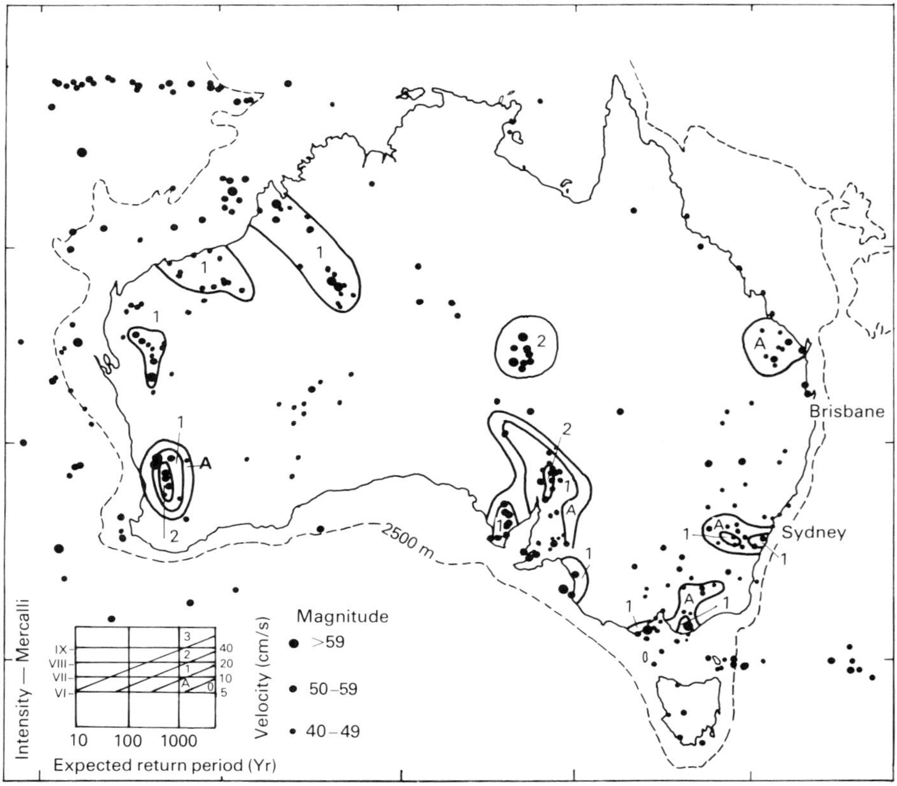

Fig. 10.7 Earthquake hazard map for the continent of Australia.

The physico-mathematical approach to hazard event prediction involves the simplification of natural processes to systems that follow physical laws and which can be described mathematically. An example of this is the soil mechanics method for the description of landslides; in this case, the factor of safety calculated is a statement of the stability of the physical system. It must always be remembered that the physico-mathematical model is constructed on the basis of a number of assumptions which may be valid, but often are not. This places a definite restriction on the method as a reliable predictor of hazard events.

A more reliable method is a statistical evaluation of recorded events, for the prediction of discrete hazards (e.g. floods, landslides, earthquakes, volcanic eruptions) as distinct from non-discrete hazards (e.g. river bank erosion, slope creep, ground subsidence). The basis of this method is the fact that discrete hazards tend to be repetitive in time. The raw data for the analysis consist of records of times of events, and intensity of events, from which the probability of recurrence of an event of a given magnitude in time can be calculated. In the case of non-discrete hazards, the rate at which the event is proceeding can be measured, and used to predict future development. In many cases, rates may vary seasonally, so that variation in rates has to be taken into account. For example, with stream bank erosion, it can be measured that erosion is occurring at a rate of x m/yr. If it is intended to construct, say, a building near the stream, without any remedial measures being taken, and the building is to have an economic life of 25 years, then the building would need to be sited at a distance from the stream of more than $25x$ m.

Most natural hazard processes are very complex, and difficult to model accurately, or analyse statistically. A lack of measurements or observations through time may add to the difficulty. Nonetheless, the engineer may have to make practical decisions. In such cases, observations can be made of the present situation, and deductions made from the results of the observations. As an example, a geologist may observe tension cracks on a slope, with groundwater seepage along the slope. The deduction from this is that the slope is unstable, and likely to fail. This method can be extended by the use of aerial photographs taken at intervals over a period of years, old maps, historical records, and personal interviews with residents.

Engineers and engineering geologists involved in hazard prediction tend to favour the observation–deduction and the statistical methods. In some cases, such as earthquake and volcanic eruptions, geophysical techniques are being employed to gather data which will increase the reliability of prediction of hazard events. Similarly, borehole tilt meters are being used to indicate slope movements. At present, however, it is possible only to predict the probability of a natural hazard event: it is not possible to predict accurately, if at all, when the event will occur.

10.6 Environmental impact statements

Environmental impact can be defined as the actions produced by a development, or by works which are capable of affecting the characteristics of the environment, i.e. of the area involved in, and surrounding, the development or works. Laws are now in force in many countries which enforce the investigation and evaluation of the consequences to the environment of all large engineering works such as dams, pipelines, motorways, hydroelectric schemes and mining projects, and of urban development. Geologists are invariably involved in environmental impact investigations. They are normally the first in the field, and are in a unique position to raise possible issues which the office investigator may not be aware of. It must be borne in mind that not all impacts are necessarily harmful. Some effects may be quite beneficial. It is obvious that the subjective element is likely to be dominant in an environmental impact study. Leopold *et al* (1971) have devised a matrix system of analysis which tends to make the assessment more objective. This approach aims to solve some of the basic problems of impact studies — the quantification of data, the establishment of cause–effect relationships and the weighting of impact magnitudes.

The environmental impact studies for large dam projects and mining projects can be used as illustrations of the issues involved in the preparation of environmental impact statements. Impacts have to be considered at two scales, the regional and local. At the regional level, beneficial aspects can accrue — flood control (e.g. Brisbane River, Queensland, Australia), hydroelectric power generation, water supply for irrigation and domestic purposes. The Dartmouth Dam in Victoria, described earlier, was constructed to reduce salinity in the Murray River. Disadvantages which could result from a large dam are biological: alteration of the downstream (and possibly upstream) ecosystem; medical: the possible increase in disease vectors such as mosquitoes; sociological: the loss of land and the alteration of microclimates; and aesthetic: the loss of the original valley.

As an example, the High Aswan Dam in Africa may be considered. The dam provides benefits in increased irrigation potential and an increase in the fishing industry. However, deleterious impacts have been beach erosion, and the cessation of beach-nourishment material supplies; and an increase in disease vectors especially mosquitoes, the buffalo gnat and the bilharzia snail. In

Fig. 10.8 Surface subsidence over coal mines east of Cessnock, New South Wales, Australia. (Photograph by courtesy of R. W. Corkery.)

Fig. 10.9 Mineral sand mining, Ballina, New South Wales, Australia. (Photograph by courtesy of M. J. Knight.)

Fig. 10.10 Tin mining at Emmaville, Queensland, Australia. (Photograph by courtesy of R. W. Corkery.)

Fig. 10.11 Overburden wastes from brick pits Liverpool, New South Wales, Australia. (Photograph by courtesy of R. W. Corkery.)

addition, numerous water weeds, in particular the water hyacinth, are flourishing and creating problems.

It is at the local level that geology becomes more directly involved. Local effects of a large dam include soil erosion, especially during the construction period; river bed erosion downstream; siltation in the storage; earthquake initiation from the water load; and changes in the groundwater regime. The latter can have a profound influence on slope stability. Local changes of less direct interest to the geologist include changes in the fauna and flora; and pollution generally during construction.

The impact of mining on the environment has to be considered in terms of the several phases of the project, and on the type of mining. In the first of these the impact of exploration activities has to be considered as well as that of mine development. In the second, the impact of open-cut mining may differ from that of conventional underground mines, while the effects of beach sand mining demand special consideration. The product mined has to be taken into account; the impact of coal mining may be quite different from that of metalliferous mining, while in the latter case the type of metal being mined may be quite significant.

The exploration phase, as a general rule, has little impact on the environment. Remote sensing techniques have none, while geological field surveys result in less damage than recreational walking. Nonetheless, the passage of four-wheel drive vehicles and the use of explosives in seismic prospecting can result in damage to flora, and possibly constitute a locus for the initiation of soil erosion. The real need for study at this phase centres on the assessment of soil erosion potential, and, if this points to possible dangers, rules must be laid down, and observed, which regulate erosion-inducing activities.

Significant environmental impacts can occur at the late stage of exploration in connection with the drilling of sites and detailed geophysical and geochemical exploration. The construction of roads to give access, and service, to sites constitutes a permanent effect on the environment; the actual area involved may be small, but the aesthetic damage can be considerable. Road construction works can also cause the spread of plant diseases, an impact noted in south-western Western Australia where jarrah dieback was spread via soil on vehicle tyres. At investigation sites, the effects of camp wastes have to be studied, as well as the effects of toxic minerals in drilling slurries. These should be confined.

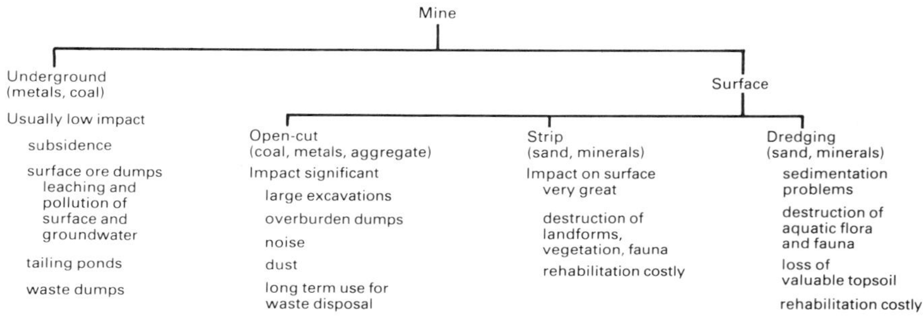

Fig. 10.12 Environmental impact of mine development.

The effects of mine development are summarized on Fig. 10.12. Underground mine development has minimal environmental impact; the most serious hazards are subsidence, which in *general*, is associated only with coal mines; and surface waste dumps. These latter can result in dust problems; safety problems (e.g. the Aberfan disaster) and the pollution of streams and groundwater by leaching from dumps and tailing ponds containing metalliferous minerals. In the case of open-cut mines, the environmental damage is more widespread and more significant. Large excavations are produced, and overburden dumps are often much larger than normal mine waste dumps. In addition, crustal readjustments around very large excavations can result in damage to buildings and services. This has occurred, for example, in a zone up to 1 km wide, around the Latrobe Valley open-cut brown-coal mines in Victoria, Australia. Furthermore, environmental damage by dust and noise can be quite serious. Open-cut mines, however, have potential for rehabilitation as lakes combined with dump landscaping, for recreation purposes, or alternatively, as sites for waste disposal.

Strip mining is particularly damaging: the impact on the land surface is severe; landforms are destroyed or seriously modified, and the natural vegetation is completely destroyed. As a result, the native fauna is lost to the area, often permanently, and although rehabilitation can be attempted, this is a very long term process and there is no clear evidence that the attempts are completely successful. In any case, rehabilitation is very costly. Attempts to rehabilitate the

Myall Lakes sand mining area in New South Wales, Australia, have, some 10 years later, been only marginally successful. Dredging, usually in river valleys, has equally serious effects, but these are often much more widespread than is the case for strip mining. Sedimentation problems can arise downstream of the dredging site and the stream flow regime possibly disastrously affected. Valuable topsoil, essential for agricultural purposes is lost, while aquatic fauna and flora are destroyed, often completely. Rehabilitation is extremely costly, and unless great care has been taken to preserve topsoil, replacement is rarely economically possible. Even if this is the case, the natural soil profile has been destroyed, and the land, even after rehabilitation, remains either poorly productive or unproductive. This has proved to be the case in the Ovens Valley of Victoria after gold dredging, and in parts of Malaysia after tin dredging.

The environmental impact statement for mining has to consider not only the actual effects on the environment of exploration and development; frequently the possibility of rehabilitation, and the costs of rehabilitation have to be considered. Rehabilitation costs possibly can be such as to render the mining operation uneconomical. It is imperative that this be known *before* mining commences. The environmental impact study becomes, therefore, not only desirable, but essential.

The geologist is only one member of a multidisciplinary team involved in the preparation of an environmental impact statement. His task is to gather relevant geological data and to analyse these data to assess the impact of the proposed works on the geological environment insofar as this affects the total environment. The geologist's role does not stop at this stage. He has to give consideration to, and make recommendations for, means by which environmental impacts can be monitored and, if necessary, corrected.

10.7 Earthquake hazards and nuclear power stations

The widespread development of nuclear power stations, and the possibility of accidents occurring which could result in disastrous effects on the environment, including people living in the vicinity, require a very careful consideration of the earthquake hazards at the sites of these stations. Since those responsible for the location and design of nuclear power stations are aware of the dangers, a great deal of work has been done to develop reliable methods of hazard prediction, and also to develop methods of analysis as a basis of design of the power plants.

Nuclear power stations are designed to withstand the effects of two possible earthquake events. These are the 'operating basis earthquake', which is the maximum level of ground motion which could be expected to occur during the life of the plant (normally about 40 years); and the 'safe shutdown earthquake', which is the maximum level of ground motion that could possibly occur at any time in the future. The structures and equipment must be designed so that operations can continue, with safety, during an operating base earthquake. Similarly designs must be such that the plant can be shut down safely in the event of a safe shutdown earthquake.

Methods of analysis are quite complex, but the basic data required include geological and soils information, as well as data on past earthquake events. The response of a structure to earthquake vibrations is influenced by the deformability, and the natural frequency, of the foundation rocks and soils. The foundation–structure interaction is of fundamental importance in the design of earthquake resistant structures. The more significant factors in the foundation material–structure interaction are the non-linear deformation behaviour of the soil and rock; the shape, size, flexibility and depth of the foundations; and the angle of incidence of the vibrations generated by the earthquake. Ideally, the foundations should be placed on rock; the hazard increases significantly if soil foundations are adopted. Despite all of the investigations, and analyses, the most important factor to be considered remains the seismic history of the area in which the nuclear plant is to be sited: the frequency, intensity, magnitude and ground accelerations of past earthquakes are essential information.

References and further reading

Anon. (1975) The EIS Technique. Australian Conservation Foundation.

Anon. (1975) Desert sites for radioactive waste graveyards. *California Geology* **28**,

Bolt B. A., Horn W. L., Macdonald G. A. & Scott R. F. (1975) *Geological Hazards*. Springer Verlag, Vienna.

Collins J. C. (ed.) (1960) *Radioactive Wastes, Their Treatment and Disposal*. Spon, London.

Cook T. D. (1972) Symposium on underground waste management and environmental implications. *Am. Assoc. Pet. Geol. Mem.* 18.

Coombs D. S. (1981) East Abbotsford landslide. *Bulletin de liason des laboratoires Ponts et Chausees Special X*. Jan. 1981.

Flawn P. T. (1975) *Environmental Geology*. Harper, New York.

Galley J. F. (1968) Sub-surface disposal in geologic basins. *Am. Ass. Pet. Geol. Mem.* 10.

Keller E. A. (1981) *Environmental Geology*. Charles and Merrill, Columbus.

Knight M. J., Leonard J. G. & Whiteley R. J. (1978) Lucas Heights waste landfill and downstream leachate transport. *Bull. Int. Assoc. Eng. Geol.* **18**, 45–64.

Le Grand H. F. (1964) System for evaluation of contamination potential at some waste dispoal sites. *J. Am. Water Res. Assoc.* **56**, 959.

Leopold L. B., Clarke F. E., Hanshaw B. B. & Balsley J. R. (1971) A procedure for evaluating environmental impact. United States Dept. of Interior, Geological Survey Circular 745.

Tank R. W. (ed.) (1973) *Focus on Environmental Geology*. Oxford University Press, New York.

Appendix 1

Geological column and time scale

Era	System or Period	Epoch	Approximate age in millions of years of base of system
	Quaternary	Recent	
		Pleistocene	1
		Pliocene	
		Miocene	
Cainozoic (recent life)	Tertiary	Oligocene	
		Eocene	
		Paleocene	65
	Cretaceous		135
Mesozoic (middle life)	Jurassic		180
	Triassic		225
	Permian		280
	Carboniferous		345
	Devonian		395
Paleozoic (ancient life)	Silurian		430
	Ordovician		500
	Cambrian		570
Precambrian	Many local systems and series are recognized		

Appendix 2

Properties and identification of minerals

Properties of minerals which aid in identification are: hardness, specific gravity, crystal habit, colour, lustre, cleavage, and fracture; and for the ore-minerals, streak. Hardness is defined by an arbitrary scale of standard minerals, known as Mohs' Scale of Hardness, arranged in degrees of increasing hardness from 1 to 10 (Table A2.1).

Table A2.1 Mohs' Scale of Hardness.

Hardness	Standard mineral	Scratched by		
10	Diamond			
9	Corundum			
8	Topaz			
7	Quartz			file
6	Feldspar			↑
5	Apatite		↑	knife
4	Fluorspar	↑	copper	|
3	Calcite	finger	coin	|
2	Gypsum	nail	|	|
1	Talc	|	|	|

The specific gravity of a mineral can be determined quite accurately in the laboratory. In the field, this is not possible, and it is usual to describe specific gravity as 'light', 'medium' or 'heavy'.

Crystal habit is determined by the physical conditions at the time of formation and growth of the mineral. Since the mode and conditions of formation of a mineral may sometimes be an indication of what the mineral is, the shape of the mineral, or its crystal habit, may be helpful in the identification of the mineral. Standard descriptive terms for crystal habit are shown in Table A2.2.

Table A2.2 Crystal habit.

Individual crystals	
Platy or tabular	One pair of parallel faces much larger than the others
Prismatic	Crystal is elongated in one direction
Acicular	Crystal is very long and needle-like
Fibrous	Long fibrous crystals.
Crystalline aggregates	
Dendritic	Crystals diverge from each other like branches
Reniform	Kidney-shaped
Botryoidal	'Bunch of grapes' habit
Amygdaloidal	Infilling of vesicles

Colour, lustre and streak can be considered together. The colour of a mineral is that of its surface, and is often controlled by impurities which are present. Streak is the colour of the powdered mineral, and is usually determined by rubbing the mineral on a white, unglazed porcelain tile. Streak has no real diagnostic value for light coloured minerals, and finds its greatest use in the determination of the dark ore minerals. Lustre is the description of the way in which light is reflected from the surface of the mineral. Its diagnostic value is not great. Descriptive terms are shown in Table A2.3.

Table A2.3 Lustre of minerals.

Metallic	Lustre of polished metal
Submetallic	Less brilliant, somewhat dull
Dull	
Vitreous	Glassy
Resinous	Oily
Satiny	
Dull	

In many minerals, planes of weakness exist, a function of ionic structure, along which the mineral will split, or cleave, more or less readily. These planes are termed mineral cleavage. Within a mineral there may be one, two, or three sets of cleavage, or none. The cleavage may be perfectly or imperfectly developed. The surface along which a mineral breaks, which is not a cleavage plane, is termed the fracture. In an uncleaved mineral, such as quartz, fracture is usually conchoidal, or shell-like, but fracture may be even, irregular, or hackly.

Table A2.4 lists the diagnostic properties of some common minerals. It should be noted that, in identifying a mineral using the properties described, all of the properties need to be considered. It is certainly unwise to attempt identification on the basis of one or two of the diagnostic properties. A number of properties such as transparency, malleability and tenacity, not generally of importance, may be valuable diagnostically in special cases.

Table A2.4 Properties of some common minerals.

Mineral	Colour	Lustre	Streak	H	SG	Cleavage	Fracture	Remarks
Quartz	Colourless	Vitreous	White	7	1.6	nil	Conchoidal	Colour variable from colourless to pink, grey, mauve
Feldspars	Variable	Vitreous	White	6	2.7	2	On cleavages	Colour white, pink, green. Cleavage usually perfect
Augite	Black	Vitreous	White	5½	3.3	2	On cleavages	Cleavages perfect. Intersect at $90^0\pm$
Hornblende	Black	Vitreous	White	5½	3.3	2	On cleavages	Cleavages perfect. Intersect at $60^0/120^0$
Biotite	Brown-black	Vitreous	White	2½	3.5	1	On cleavage	Cleavage perfect
Muscovite	White	Vitreous	White	2½	2.7	1	On cleavage	Cleavage perfect
Garnet	Red	Vitreous	White	7	3.6	nil	Uneven	Colour varies
Olivine	Green	Vitreous	White	6½	3.5^+	nil	Uneven	
Calcite	White	Vitreous	White	3	2.7	3	On cleavage	Three perfect rhombohedral cleavages. Mineral effervesces in HCl.
Gypsum	White	Satiny	White	2	2.3	3	On cleavages	Cleavage perfect in one direction, two less perfect.
Dolomite	White	Dull	White	3½–4	2.9	3	On cleavages	Three perfect rhombohedral cleavages
Galena	Grey	Metallic	Black	2½	7.5	3	On cleavages	Three perfect cubic cleavages.
Sphalerite	Red-brown	Vitreous	Cream	4	4	3	On cleavages	Smell of H_2S when rubbed on streak plate.
Pyrite	Brassy	Metallic	Black	6	5	nil	Uneven	
Chalcopyrite	Brassy	Metallic	Black	3½–4	4.2	1	Uneven	
Magnetite	Black	Dull	Black	5	4	nil	Uneven	Magnetic
Limonite	Orange-red	Earthy	Brown	5	4	nil	Uneven	
Hematite	Deep red to steel grey	Submetallic to Metallic	Red	5–6	5.3	nil	Uneven	

Appendix 3

Field identification of rocks

After Krynine D. P. and Judd W. R. (1957) *Principles of Engineering Geology and Geotechnics.* McGraw-Hill, New York.

Grains or crystals visible to naked eye, colour light

Angular particles				Rounded particles			Erratic large grains	
Large	Fine to medium	Very fine	Foliated or banded	Large	Fine to medium	Very fine	Rounded	Angular
Pegmatite	Tuff	Rhyolite (+Q) and Trachyte (-Q)	Schist, Gneiss	Conglomerate (+10% of grains over 2mm dia.)	Quartzite	Siltstone	Depositional breccia	Volcanic breccia and agglomerate fault breccia
Granite (+Q+F) Granodiorite(+Q+F) Monzonite (-Q+F) Syenite (-Q+F) Marble (reacts with HC1) Arkose				Sandstone (bedded)				

No grains or sparse crystals visible to naked eye

Breaks with rough surface					Breaks with smooth or platy surface (usually has bedding)				
Glassy lustre	Dull lustre	Shiny lustre	Earthy appearance		Laminated		Not laminated		
Quartzite	Rhyolite, Trachyte	Schist (foliated)	Pumice Volcanic ash	Limestone	Shale	Shale Slate Phyllite	Claystone Mudstone Serpentine (usually greasy and may be banded)	Limestone	Dolomite

Grains or crystals visible to naked eye, colour dark

Angular particles		Rounded to subangular particles
Fine to medium	Very fine to glassy	Graywacke Sandstone
Peridotite (-Q-B) Gabbro (-Q-B) Diorite (-Q+B) Dolerite	Andesite Basalt	

No grains or sparse crystals visible to naked eye

Glassy lustre		Laminated	Non-laminated
Obsidian	Shale	Shale (flexible) Slate (brittle) (dull) Phyllite (shiny)	Basalt Mudstone or claystone

Appendix 4

Stereographic and equal area projections

Stereographic and equal area projections provide techniques for the graphical representation and analysis of planar and linear defects in rocks. All planar defects have an orientation in space which can be defined by two parameters: dip and strike. The strike of the plane is the bearing of a horizontal line in that plane; the dip is the inclination of the plane, from the horizontal, in a direction normal to the strike. Thus, we can speak of the orientation of a joint as being strike 300° (or alternatively N60° N, 120° or S60° E) and dip 50° SW (or 50° S 30° W). The stereographic projection of this plane is a two-dimensional representation of the three-dimensional orientation. If it is imagined that the structural plane lies in a sphere, then the intersection of the plane with the surface of the sphere will be a great circle. The spherical projection of that part of the great circle lying in the lower hemisphere, onto the equatorial plane, is the stereographic projection of the structural plane.

STEREOGRAPHIC PROJECTIONS

For the plotting of stereographic projections, a stereographic, or Wulff, net is used (Fig. A4.1). The procedure is simple, and consists of the following steps:

(1) Place a thumb tack through the centre of the net, and then place a piece of tracing paper over the net, with thumb tack penetrating the centre of the paper. Mark North point of net on tracing paper. Trace perimeter of net.

(2) Mark on tracing paper, on the perimeter, the strike direction of the plane to be plotted.

(3) Rotate tracing paper until strike mark coincides with North point of net.

(4) Along the East–West diameter of the net, measure off the dip of the plane, in the appropriate direction. The East and West points on the perimeter correspond to 0° dip; the centre of the net to 90° dip.

(5) Trace out the great circle joining the strike point and the dip point. This is the stereographic projection of the plane.

(6) Rotate tracing paper until North point coincides with the North point of the net.

It will be appreciated that the orientation of a structural line, including the line of intersection of two planes, can be determined by stereographic projection. The orientation of a line is defined by its trend and plunge. The trend is the strike

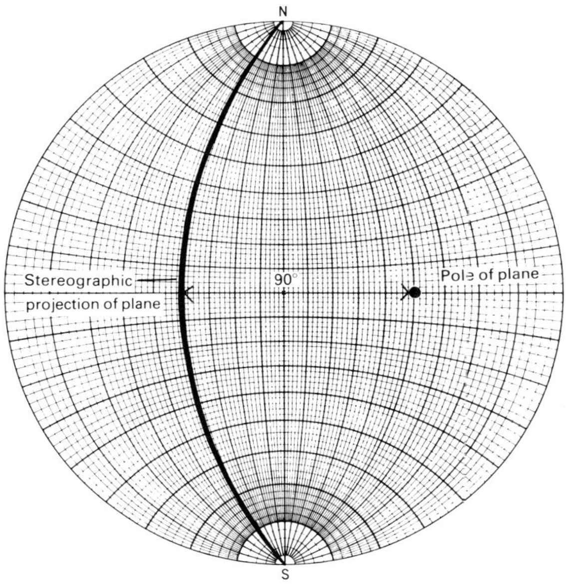

Fig. A4.1

of a vertical plane containing the line, and the plunge is the inclination, from the horizontal, of the line in the vertical plane.

To determine the trend and plunge of a line (e.g. the line of intersection of the two planes of Fig. A4.1), the following procedure is adopted:

(1) Rule a straight line from the centre of the net through the point of intersection of the two planes. This is a stereographic projection of the vertical plane containing the line.

(2) Read off the strike of this plane. This gives the trend of the line.

(3) Rotate the tracing paper until the straight line ruled lies on the N–S (or E–W) axis of the projection.

(4) Measure the plunge along this axis. Zero plunge is at the perimeter.

To plot the projection of a line:

(1) At the perimeter of the net, mark on the tracing paper, the trend of the line.

(2) Rotate the tracing paper until the point marked coincides with the N or S points of the net.

(3) Measure off the amount of plunge along the N–S axis, and mark the point. This is the stereographic projection of the line.

POLAR AND EQUAL AREA PROJECTIONS

When it is necessary to consider a large number of planes, it is convenient to plot the projection of *pole* to each plane, rather than the stereographic projection of the plane itself. The pole to a plane is the normal to that plane (Fig. A4.2). For the statistical analysis of a large number of planes, poles to the planes are plotted

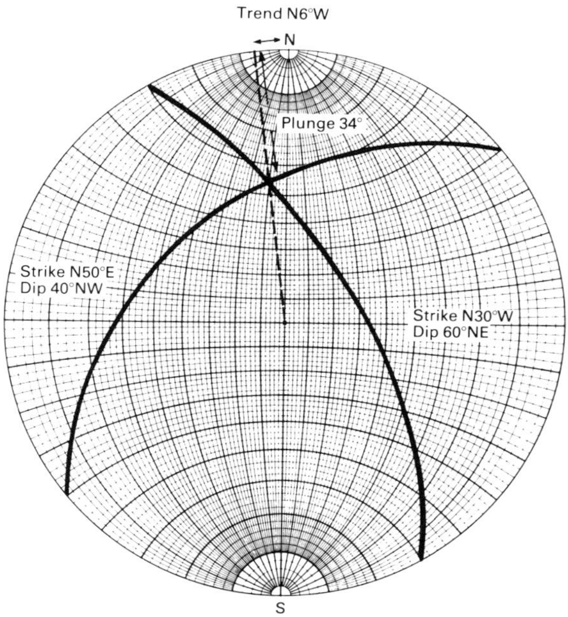

Fig. A4.2

on the equal area, or Schmidt, net (Fig. A4.3). It is possible, often by visual inspection alone, to determine from such a plot whether or not a definite pattern exists. More accurate results are obtained by counting the intensity of poles, usually expressed as a percentage of the total number of poles per 1% of the area of the projection. To emphasize the pattern, the diagram is contoured with contours joining points of 1, 2, 3... % of points per 1% unit area of projection.

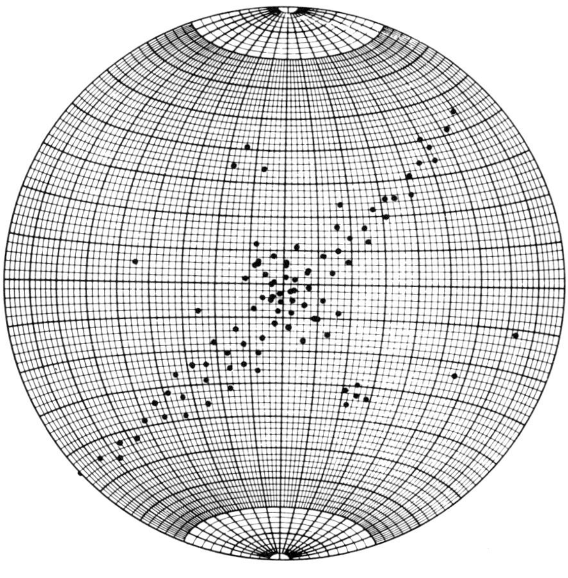

Fig. A4.3

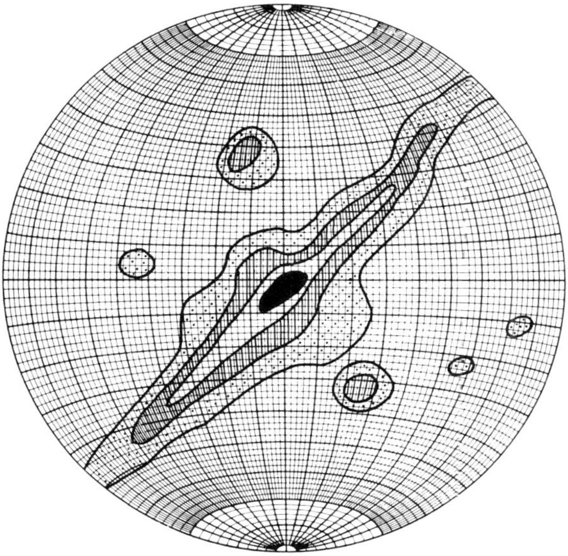

Fig. A4.4

This procedure can be performed manually; this, however, is both tedious and time consuming. It is quicker to make use of computer techniques, whereby the raw data (dip, strike) are supplied to the computer, programmed to produce the contoured diagram.

If a pattern exists, the contoured diagram will have well defined maxima. The centre of each of the maxima can be regarded as the pole to the mean of the set of planes which contributed to a particular maximum. From this it is possible to derive the pattern of fractures present in any given section of a rock mass (Fig. A4.4). The equal area projection is used not only for the analysis of macrofractures in a rock mass, but may also be used for the analysis of microfractures in the rock material, such as grain boundaries, and cleavage planes. In this case, the orientation of the planes is measured microscopically on orientated thin sections using a universal stage.

One very important use of the equal area projection is the determination of the symmetry of the rock fabric. The symmetry of the diagram is that of the particular fabric under examination.

Author Index

Index of Rocks, Minerals and Soils

General Index